The Food Debate:
A Balanced Approach

R. K. Proudlove, BSc, MPhil, FIFST
Principal Lecturer in Food Studies,
Humberside College of Higher Education, Grimsby

Hodder & Stoughton
LONDON SYDNEY AUCKLAND TORONTO

The author and publishers would like to thank the following for permission to reproduce material in this book:

Prof. J. H. Freer, University of Glasgow (page 56), Golden Wonder Ltd (page 46), M. Grebby (pages 4, 20, 30, 44), H. J. Heinz Company Ltd (page 36), International Coffee Organisation (page 148), McVities (pages 32, 136), D. M. Marley, F. Duerr & Sons Ltd (page 123), Meat and Livestock Commission (pages 106, 108), Milk Marketing Board (page 78), National Dairy Council (pages 41, 71, 76, 84), National Osteoporosis Society (page 29), Sea Fish Industry Authority (page 115), Van den Berghs and Jurgens Ltd (page 93), and The Minisistry of Agriculture, Fisheries and Foods for permission to use the list of permitted additives to be found in their booklet Food Additives – the Balanced Approach (HMSO, London, © Crown Copyright 1987) as the basis for the list found in the Appendix; and also for permission to redraw the diagram of a food label found in the same publication.

British Library Cataloguing in Publication Data
Proudlove, R. Keith
   The food debate.
   1. Food
   I. Title
   641.3

   ISBN 0–340–52748–X

First published 1990

© 1990 Keith Proudlove

Typeset by Wearside Tradespools, Fulwell, Sunderland
Printed in Great Britain for Hodder and Stoughton Educational, a division of Hodder and Stoughton Ltd, Mill Road, Dunton Green, Sevenoaks, Kent by St Edmundsbury Press Ltd, Bury St Edmunds, Suffolk

Contents

Introduction

The subject of food appears in the press and on television at least once a week. There are many views expressed by so-called 'experts' on current topics such as the use of additives, irradiation and nutrition. Many of these views are biased, inaccurate and are often from people with vested interests elsewhere. This book has been written to try to give a balanced view of current issues, and to give an accurate understanding of foods, their composition, nutritive aspects and effects of cooking.

The book is divided into three main sections: Section I covers the current issues relating to food; Section II is a review of a range of foods commonly found in our diet; Section III is a glossary of words and terms commonly used when discussing the science and technology of foods. In the Appendix, a complete list is given of permitted additives, their uses and E numbers.

This book is suitable for a wide range of Food Science, Catering and Home Economics courses, from A-Level to BTEC Higher and Degree levels. Parts of the book are also suitable for GCSE Home Economics, Food Studies and Chemistry or Biology options. It is also designed to be understood by the informed general reader.

Note: In Sections I and II words that appear in the Glossary are set in bold type the first time they are used in the text.

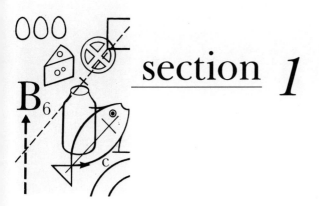

section *1*

Current
Issues

1 Additives: a balanced view

Much has been written about **additives**, but little has been said in support of their use. To many people, additives are both unnecessary and harmful, with the letter *E* characterizing this distrust. And yet, additives have allowed the production of foods of predictable shelf-life, of high nutritional value and of infinite variety. Why then have these substances received so much interest in the media?

It has been good press to criticize the Food Industry and attack the composition of modern manufactured foods. The Industry is a diverse one (with interests ranging from meat production to soft drinks) and is unable to speak as one body. Some additives, particularly colours, have had dubious beginnings. Colour was used during the last century to disguise changes, often spoilage, in foods transported to the cities. Poisonous colour compounds were often used, especially from the textile industry, containing arsenic, lead, mercury and copper. More recently the colour **tartrazine**, *E102*, has been found to cause allergic reactions in some people and even hyperactivity in some children. Unfortunately, although not acceptable, many people extrapolate this allergic nature of tartrazine to include all other colours and even all other additives. It is estimated that about 1 person in 10 000 is allergic to tartrazine, but few people point out that about 9 per cent of the population is allergic to the protein of milk! However, in spite of the inaccuracies of the media, the Food Industry has been quick to respond to public reaction by removing *E102* from its products.

The modern Food Industry is, in effect, a fashion industry with its own trends and fads. If the public, even for dubious reasons, does not want additives the Industry will react to this need and develop additive-free products. However, problems during manufacture, with storage, nutrition and safety may well arise. For example, for many years meats have been cured with nitrates and nitrites in various combinations. Evidence was gradually accumulated that dangerous carcinogenic substances called **nitro-**

3

samines could be produced by these curing salts in such products as bacon. As a result, levels of nitrates and nitrites have been significantly reduced. But nitrite is effective in destroying a very dangerous bacterium called *Clostridium botulinum* which produces a toxin with a 70 per cent mortality rate. The risk of this bacterium causing illness must therefore increase as curing salts are used in lower amounts. Fortunately, this risk is still very slight, but so is the risk of cancer from eating the carcinogenic nitrosamines. The latter risk has been estimated as being considerably less than being killed by walking in front of a double-decker bus!

Products containing additives

The Food Industry has been accused of using poor quality raw materials and of masking this poor quality by the use of a range of additives. Anyone who has been round a food factory will know that very strict quality control procedures and specifications are applied to raw materials. Food processing, at best, only maintains the quality of the raw material, but very often there is a slight drop in quality. So poor quality raw materials lead to low quality manufactured or processed foods, which the public will not buy. This is something rarely seen in Britain, but, if it does happen, the firm will likely attract prosecution under the various Food Acts.

The levels of hygiene used in some factories are alike to those used in operating theatres, because manufacturers and employees realize that, if their products cause any disease, then very quickly they will be calling in the receiver. Examples have occurred to illustrate this over the last few years:

- In the Sixties the Aberdeen typhoid outbreak was caused by Argentinian corned-beef, the sales of which ceased and have never recovered

- The Canadian salmon industry was involved in a small out-break of botulism, the disease caused by the toxin produced by *Clostridium botulinum,* which resulted in lost sales in the region of $300 000 000

- In 1986 a well-known baby-food manufacturer went into liquidation after causing just one outbreak of **salmonella** food poisoning

So, is a food manufacturer going to put a chemical into a food product that will cause illness in any form when faced with consequences like these?

The anti-additive lobby is so strong that the food products we eat are being forced into rapid change. The words 'natural' or 'nature identical' appear on many products and 'natural' ingre-dients are used increasingly as opposed to artificial ones. Any-thing natural must be 'good' and anything 'artificial' or 'synthetic' must be 'bad'. Yet, some of the deadliest poisons are natural. Many substances occur naturally in foods which block the work of our digestive **enzymes**, bind up our **vitamins** or essential miner-als, increase or lower our blood pressure or irritate our stomach lining. Some of these substances are far worse than any permitted additive used by the Food Industry.

A food manufacturer will only use an additive if no other way is possible to solve a particular problem. A manufacturer should not use an additive just because it is legally permitted. Scientific opinion develops and changes and, eventually, with it so do the regulations. The overall wholesomeness of a food is vitally impor-tant; it is an excellent selling feature for a manufacturer, and thus it should not be put in jeopardy.

Food is an excellent medium for the growth of micro-organisms so it will quickly lose its wholesomeness. Preservatives have been used as food additives to extend the storage life of a product and to keep it safe from the growth of harmful micro-organisms, particularly pathogenic bacteria. Many products now boast a 'preservative-free' label; often **preservatives** were never used in

5

them in the first place! A bottle of orange squash can be opened and used over quite a long period, with the help of a refrigerator for several weeks. The preservative used here, **benzoic acid** (*E210*) or one of its derivatives, has been implicated in allergic reactions for some people, but we are all subject to the effect of pathogenic bacteria!

I hope I have managed to convey that to ban additives would be a foolish step; they must continue to be used in a controlled and sensible way.

What are additives?

Having discussed the arguments for and against additives now let us turn to the actual nature and function of food additives.

Additives are chemicals, both synthetic and natural, that are used to give various functional properties to foods. The additives, in the quantities used, are edible but are not foods in their own right. Some additives are very widespread in nature (for example **pectin, ascorbic acid** (vitamin C)), others come from specific sources, (for example **gums** from certain seaweeds). Naturally-occurring additives must be treated and controlled in exactly the same way as synthetic ones.

Functional properties of additives

It is possible to divide additives into four main groups according to their functional properties. Additives in a food alter either:

(a) its physical characteristics

(b) its sensory characteristics, such as flavour, texture and colour

(c) its storage life

(d) its nutritional status

In addition to these four, another group of additives is used by manufacturers as an aid to the production of large quantities of high quality foods. In fact, some additives can fulfil more than one function. For example thickeners such as pectin fall in groups (a) and (b). Vitamin C (ascorbic acid) is an **antioxidant**, therefore belonging to group (c), but is also a vitamin, belonging to group (d).

(a) Additives affecting the physical characteristics of a food

A food product may be:

- made thicker
- made more or less acid
- aerated with gas bubbles
- emulsified.

To make a product satisfactory during consumption it must have the right 'mouthfeel'. A number of additives are available which give body to a product. Gravies and syrups are thickened with **starch** or flour. Sauces may have small amounts of gums added. Instant desserts rely on specially formulated starches (**pre-gelatinized**) which thicken with the addition of cold water or milk.

Common additives in this group include:

Carageenan (*E407*) – used in milk shakes and desserts
Gum arabic (*E414*) – used in confectionery
Pectin (*E440 a* and *b*) – used in preserves
Agar (*E406*) – used in ice-cream

Some products are too acid for most people whereas many products need to be made more acid to be acceptable. To control the acidity of a product such as sugar confectionery a substance known as a **buffer** may be used. Many types of **acids** are used to increase the acidity of a product, common examples being:

Sodium lactate (*E325*) – a buffer used in preserves and confectionery
Calcium citrate (*E333*) – a buffer used in soft drinks, sweets and preserves
Acetic acid (*E260*) – vinegar used in pickles, mayonnaise
Tartaric acid (*E334*) – acid used in desserts and raising agents
Citric acid (*E330*) – widely used acid, useful in binding up traces of metals

The use of raising agents has made possible a vast range of baked products. Sodium hydrogen carbonate (bicarbonate) and an acid such as cream of tartar (*E336*) or calcium hydrogen phosphate (*E341*) make up the raising agents. If the additive has been approved by the EC it has an *E* in front of it. Some additives have a number but no letter *E* as yet. Sodium hydrogen carbonate is an example, surprisingly, with the number *500*.

Emulsifiers and stabilizers play vital roles in many food products; without them the food becomes unstable and separates out into watery and fatty layers. An **emulsifier** allows the dispersion of tiny droplets of oil to be made in water to give a stable emulsion. Examples occur in mayonnaise, sauces, drinks and soups. A **stabilizer** usually works by absorbing large quantities of water and binding them into a stable form. In ice-cream a stabilizer is added to prevent the formation of large 'crunchy', ice crystals during the freezing process. The ice-cream will thaw gradually with a stabilizer present and not drip over the hand as sometimes happens with home-made products.

Common emulsifiers include:

Monoglycerides of fatty acids (*E471*) – found in frozen desserts
Sucrose esters of fatty acids (*E473*)   – used as a wetting agent

Stabilizers include those listed above to give body to a product and:

Sodium alginate (*E401*) – used in cake mixes
Guar gum (*E412*)   – found in packet soups
Xanthan gum (*E415*)   – used in pickles and coleslaw

In general, to act as an emulsifier part of the structure of a substance must be capable of dissolving in water and part in fat or oil. In this way the emulsifier arranges itself on the interface between the fat droplets and the water and thus prevents droplets joining together and rising to the surface, as cream rises in milk on standing.

Carbohydrates, particularly the starches and gums, can act as stabilizers as they have the ability to absorb water. Some of these substances often act as emulsifiers as well.

(b) Additives affecting the sensory characteristics of a food

A food, to be enjoyable, must taste, smell and feel right in the mouth. It must also be attractive to the eye. In Britain, until recently, we found coloured foods more attractive. However, there is now a trend towards a preference for non-coloured and naturally-coloured foods.

Food flavourings, although there are about 3 000 of these, are not given *E* numbers. The majority of **flavour** substances used in food are natural and, contrary to public opinion, the number of synthetic flavours is very small. Spices and herbs have been used as flavourings from the earliest times. In fact, so valuable were spices in Roman times that adulteration of a spice carried the death penalty.

As natural flavours are often limited in availability and reliability, there is considerable development taking place in 'nature-identical' flavourings. These are either synthesized or extracted from natural materials and are, in both cases, identical to substances naturally present in food materials. These flavourings are often cheaper, more readily available and behave in a predictable manner during processing and product storage.

Flavours are, in fact, odours and are detected by the nose. We taste only salt, sour, bitter and sweet. Fruit flavours are often termed **essential oils** arising from the word 'essence' and not from being essential to life. Less volatile natural flavourings are extracted by **solvents**, for example from dried spices. These flavourings may be called **oleoresins**.

A number of substances have been found to have the ability to enhance the flavour of other substances and to modify or mask undesirable flavours. The best known **flavour enhancer** is **monosodium glutamate** (MSG) (*621*). It stimulates the taste buds in the mouth. MSG has been used by the Chinese for generations and was found to be the main flavour component of soy sauce. Large amounts have been known to cause sickness or dizziness, and the sickness was given the title of **Chinese restaurant syndrome**! Manufacturers have advertised foods as MSG-free as if this is a particularly harmful additive. However, very many foods contain MSG, an **amino acid** derivative, as a natural component!

The British have a particularly sweet tooth and so sweeteners, particularly sugar, have been used extensively. Sugar usage has decreased (see Chapter 2) and the growth of artificial sweeteners has been considerable. **Saccharin** has been used for many years, but newer sweeter substances are now available. **Aspartame** (no *E* number) has in its relatively short life carved out a large niche for itself as a sweetener for soft-drinks, particularly the 'diet' variety. The sweetener is made up of two naturally-occurring amino acids, aspartic acid and phenylalanine. Unfortunately, in acid drinks, over a period of time it breakdowns and so products using it must be less acid, with a shorter shelf-life. The sweetener acesulfame is similar in sweetness but more stable.

The visual appearance of a product is a vital selling feature. Colours have been used to modify the appearance of a product to make it more attractive to the consumer. They have been used to colour a food deficient in colour due to certain processing or to ensure a consistent product colour. Recently the consumer has turned against colours more than any other additive, particularly the azo dyes, such as tartrazine (*E102*). There are 58 permitted colours and **caramel** (*E150*) is the most common. Naturally-occurring colours include a wide range of **carotenoids** (carrot-like) such as beta-carotene (*E160a*) and lycopene (*E160d*) which occurs naturally in tomatoes.

Additives which affect the mouthfeel of a product have been mentioned among group (a). Mouthfeel is a vital characteristic of a product and must be right, whether it be firmness, crispness, smoothness or chewiness. Considerable skill is required in ensuring modern manufactured food products have the desired mouthfeel.

(c) Additives affecting the storage life of a food

One of the major successes of additives such as preservatives is their ability to extend the storage life of a product over a longer period than normal. **Preservatives** help to reduce or prevent wastage of food through spoilage caused by micro-organisms. They also have the ability to help protect the public from food poisoning caused by certain bacteria. Longer shelf-life enables a greater variety of products to be kept in a store, even a small corner shop. Similarly, food can be kept in the house for longer periods and sometimes used over a period of time.

Common examples of preservatives include:

Sorbic acid (*E200*)	– used in soft drinks and processed cheese
Benzoic acid (*E210*)	– used in soft drinks
Sulphur dioxide (*E220*)	– widely used, often as sulphite (*E221–227*)
Potassium nitrate (*E252*)	– used in curing bacon, ham and other cured meats

Fats, oils and foods containing them are subject, over a period of time, to the effects of oxygen in turning the product rancid. This type of **rancidity** is accelerated by light (UV) and by certain metals, particularly copper and iron. **Antioxidants** are added to foods such as these to slow down or prevent the process of rancidity and thus extend the storage life of the product. Some antioxidants stop the chemical reactions involved in rancidity, whereas others remove oxygen from the product. Another type of rancidity, **hydrolytic rancidity**, is caused in the presence of water by some enzymes or micro-organisms. This type of rancidity, found in butter for example, is not prevented by antioxidants.

Common antioxidants include:

Ascorbic acid (*E300*)   – used in fruit drinks
Propyl gallate (*E310*)   – used in vegetable oils and chewing gum
Butylated hydroxyanisole (*E320*) – used in cheese spreads, stock cubes

Over a period of time some foods deteriorate as certain proteins and amino acids combine with certain sugars to produce brown products. This is called **non-enzymic browning** and, besides being unattractive, lowers the nutritional value of the food. Sulphur dioxide (*E220*) is used to prevent the browning.

(d) Additives affecting the nutritional status of a food

This must surely be the only group of additives never criticized by the anti-additive lobby. The group includes minerals, vitamins and protein supplements. These substances are only legally additives when they fulfil a technological purpose, for example, ascorbic acid (vitamin C) is an antioxidant. Some nutrients must be added to foods by law, for examples, vitamins A and D must be added to margarine.

Our diet receives many useful nutrients from this group, particularly in breakfast cereals. A typical flaked breakfast cereal may have the following nutrients added:

vitamin C	35.0 mg/100 g	vitamin B_6	1.8 mg/100 g
niacin	16.0 mg/100 g	vitamin D	2.8 mg/100 g
thiamine (B_1)	1.0 mg/100 g	iron	40.0 mg/100 g
riboflavin (B_2)	1.5 mg/100 g		

(e) Additives used by manufacturers during processing

Processing aids are additives that the manufacturer uses to facilitate the production of a foodstuff, usually of higher quality and often more cheaply. **Solvents** are used to extract substances from materials, for example fruit flavours from peels. **Filter aids** are used to accelerate the filtration of liquid foods in removing suspended particles. **Anti-caking agents** are added to powders to keep them free flowing, for example, magnesium carbonate (*504*) is added to ensure salt does not cake. These additives, like all additives, would not be used if an alternative method of approach to processing was available.

The safety of additives

A long process to gain approval for an additive is necessary before it can be legally used.

- Expensive research must be carried out by the additive manufacturer to ensure that it is safe
- Evidence of the additive's safety must be presented to a number of Government departments responsible for food and health
- The Food Advisory Committee, which advises the health and food ministers, decides if a new additive is necessary. The Committee must be convinced by the manufacturer that the additive gives benefit to the consumer
- The safety of the additive is examined by the Committee on Toxicity of Chemicals in Food Consumer Products and the Environment. The Committee carefully examines all research carried out on the new additive whether in the UK or abroad
- The Food Advisory Committee considers the advice on the safety of the additive and makes a recommendation to the Minister responsible as to whether the additive should be used and under what conditions

Some categories of additives must be approved by the EC, in addition to the above, and needless to say a long process of approval is required.

Ministers produce new regulations concerning the additive which are eventually made public. There is an opportunity for the manufacturers and other bodies to comment on the regulations before they are laid before Parliament.

Summary of additive numbers

When approved, an additive is given a number which is used as part of the ingredient list on a food label. A number without an *E* is controlled by the UK and not as yet by the EC.

Additives	*Numbers (in general)*
Colours	100–180
Preservatives	200–290
Antioxidants	300–321
Acids, buffers, anti-foaming agents and similar	mainly 300s, from 170 up to 900s
Emulsifiers and stabilizers	322–494
Sweeteners	420–421

The food label

It is by looking at a label that we become aware of the additives used in the product. The use of *E* numbers has highlighted additives and facilitated the anti-additive campaign. Labelling of food is controlled by a number of strict regulations so that the general public is not misled into thinking the product is something it is not.

Here we are concerned with the ingredient declaration on the label. The list of ingredients is in descending order by weight. The first ingredient listed is found in the greatest amount in the product, and the last ingredient in the smallest amount. Food materials are listed by name, for example, sugar, tomatoes, cocoa and so on. Additives are labelled by their category name, with the number identifying the substance; for example, caramel would be labelled: colour *E150*. The chemical name of the additive may also be found on some labels. If no number is available for one additive it is identified by the category name and the chemical name; for example, saccharin would be labelled: artificial sweetener – saccharin.

Acids do not need to be declared by their category group if the word 'acid' appears in their name (for example, **tartaric acid**). Flavourings are declared by the single word 'flavourings'. Water must be declared if it is 5 per cent or more of the total weight.

this food manufacturer has chosen to use the chemical names of additives instead of their serials numbers.

only the category name need be included for flavourings.

although beta-carotene is a permitted colour, and is being used as a colour in this drink, it is also a source of vitamin A.

INGREDIENTS, AFTER DILUTION:

Water, Sugar, Glucose Syrup, Comminuted oranges, Citric acid, Preservatives Sodium benzoate, Sodium metabisulphite, Artificial Sweetener (Saccharin), Vitamin C, Flavourings, Colour (Beta-carotene – Provides Vitamin A)

it is not necessary to use the category name for additives which function as 'acids' in foods and whose chemical name includes the word 'acid'.

saccharin does not have a serial number so that it must be indicated by the category 'artificial sweetener' and its chemical name.

vitamin C is the same chemical as the antioxidant E300 L – ascorbic acid. Here, however, it is being used as a vitamin.

Example of a food label from an orange drink
(Source: Food Additives – the Balanced Approach, *Ministry of Agriculture, Fisheries and Foods, HMSO, © Crown Copyright 1987)*

Examples of some ingredient lists are given below:

Example 1 – horseradish sauce

Horseradish (fresh), malt vinegar, sugar, vegetable oil, cream, salt, skimmed milk powder, acetic acid, pasteurized dried egg, mustard, stabilizers (*E412, E415*), colour (*E171*), tartaric acid, lactic acid and flavouring

The points mentioned on acids and flavouring can be seen in the list, and other ingredients are clearly understandable. The stabilizers are *E412*, guar gum and a new gum *E415*, xanthan gum. The colour used *E171* is titanium dioxide.

Example 2 – an ice-cream

Reconstituted dried skimmed milk, sugar, dextrose, vegetable oil, whey powder, emulsifier (*E471*), stabilizers (*E412, E407*), natural colours (annatto, curcumin), flavouring

The emulsifier used, *E471*, is a **monoglyceride** or a mixture of **mono-** or **di-glycerides** of fatty acids. The stabilizers prevent the formation of large ice crystals and control thawing. Here *E412*, guar gum, and *E407*, carageenan, are used.

Example 3 – a canned sauce

Tomatoes, onions, modified starch, sugar, mushrooms, salt, hydrolysed vegetable protein, vegetable oil, citric acid, spices, herbs

Apparently a 'natural product' with no *E* numbers! Modified starches are under consideration for *E* numbers in the near future.

Example 4 – mayonnaise

Water, vegetable oil (with antioxidant *E320*), modified starch, egg yolk, sugar, spirit vinegar, salt, lemon juice, stabilizers *E405*, *E415*, preservative *E202*, flavourings

This product, although traditional in many respects, will have a long, stable shelf-life. Antioxidant is contained in the vegetable oil and cannot be ignored. Here *E320*, butylated hydroxyanisole, is used. Stabilizers prevent the high level of oil from separating out, an **alginate** *E405* and xanthan gum (*E415*) are used. This product contains the preservative *E202*, potassium sorbate, which is effective against moulds in particular.

I have not been exhaustive in this treatment of the complex labelling regulations. A free Ministry of Agriculture leaflet, *Look at the Label*, is well worth reading.

Conclusion

Additives are useful aids to the processing industry and they have given us a wide range of varied, safe and innovative food products. However, they must be carefully controlled and any doubtful substances eliminated. Alternative methods to using additives must be found, but additives will remain in foods because without them many foods would disappear from the market place.

Further Reading

Ministry of Agriculture, Fisheries and Food, *Food Additives – the Balanced Approach* (London: HMSO). (Available free from FOOD SENSE, London SE99 7TT, as are MAFF's other FOOD SENSE booklets:
Food Safety – a Guide from HM Government
Look at the Label
Food Surveillance
Food and Nutrition
Pesticides and Food – a Balanced View)

Institute of Food Science and Technology, *Food Additives – the Professional and Scientific Approach* (IFST, 1986) (20 Queensbury Place, London SW7 2DR)

(See Appendix for full list of permitted additives)

2 Nutrition update

Nutrition is not a static science, it has been likened to a football game with the goal-posts continually on the move! Information given one week on a certain nutritional aspect may be contradicted the following week. At the time of writing it was reported that even a small intake of alcohol would increase the risk of breast cancer in women, one week later this was refuted. It is no wonder that many people still follow the maxim, 'a little of what you fancy does you good'. In general, it is true to say that *individual foods* are not bad; it is the *diet as a whole* that may be bad.

However, a number of trends are developing and the public is becoming more aware of so-called 'healthy eating'. In 1985, Food Policy Research at the University of Bradford, undertook research into the extent of public concern about diet:

- A surprising 95 per cent of respondents agreed that 'People are becoming a lot more conscious of what they eat nowadays'
- 50 per cent agreed that 'meat products contain more fat than is good for you'
- Some 63 per cent of respondents said they would consider altering their diet on health grounds

Yet, many of the most common and deadly diseases in the UK, and in the West in general, have a dietary basis. It would be misleading to say that a particular food *causes* a certain disease, as modern illnesses, such as heart disease, strokes and cancers, result from *extended exposure* to a number of risk factors in *susceptible people*. As well as a poor diet, factors such as lack of exercise, hereditary characteristics, smoking and too much stress contribute significantly.

Modern illness

The plagues and epidemics of centuries past are known to most people through their school history lessons, but long-term degenerative diseases affect much more of today's population. These diseases have been termed 'diseases of civilisation' and 'diseases of affluence', simply because they are rarely found in developing countries. The number one killer in Britain, heart disease, affects 40 per cent of men and 38 per cent of women. Other diseases which can be diet-related include: high blood pressure, cancers, obesity, ulcerative colitis, diverticulitis, constipation, diabetes and dental caries. A group of people taken at random will admit that they are suffering, or have suffered, from one or more of these diseases, even if it is only to having a few tooth fillings.

Heart disease

Blood vessels can become narrowed by the deposition of fatty material on the inside of their walls. The condition this produces is *atherosclerosis*, which can lead to a situation where blood clots become trapped in the narrowed blood vessels and, as a result, slow down or stop the blood flow to the heart. This causes what is known as a heart attack or 'coronary'.

A number of factors are involved including hereditary factors, stress, smoking and high blood pressure. High levels of fats and cholesterol are the main diet-related factors. It is these fatty materials which form 'plaques' leading to blocked blood vessels. **Cholesterol** is a waxy material found in animal foods only. A high intake of cholesterol in the diet does not necessarily cause an increase in blood cholesterol. The body actually manufactures cholesterol, often twice as much as it needs. In susceptible individuals, for a number of reasons, this material is deposited in blood vessels. Other more fortunate individuals never experience the problem.

Diet and cancer

There has been evidence for some time that diet could play a part in starting cancers of various types. This evidence is not conclusive and the precise components of the diet involved are still open to debate. Excess calories and excess fat have been implicated in the development of cancer of the bowel and of the gall bladder. Excess fat and meat consumption has been linked with heart cancer. In Japan, where fat intake is about a quarter of that in Western nations, the occurrence of these cancers is significantly lower. The Japanese who migrated to the United States, however, now have the same level of cancers as the Americans. Excessive salt intake may be linked to some stomach cancers, as lower salt diets have been linked to a fall in such cancer.

All is not bad, however, since there are many substances in the diet which are believed to protect against cancer. **Fibre**, vitamins A, C and E and some short chain **fatty acids**, have been shown to have some anti-cancer activity. β-carotene (a **carotenoid**) from vegetables is an example of a substance, which can be converted into vitamin A, and may have anti-cancer properties. Work continues in the hope that one day some wonder cure may be found.

Public statements and reports

In the last decade there have been a number of reports on what is a healthy diet. These reports are usually a consensus of opinion and draw upon the knowledge and experience of experts from medicine, nutrition, government and the Food Industry. These reports are given below. Of the five, the two most significant are the NACNE and COMA reports.

Eating for Health	DHSS, 1978
Proposals for Nutritional Guidelines for Health Education in Britain	National Advisory Committee on Nutritional Education (NACNE), 1983
Diet and Obesity	Royal College of Physicians, 1983
Diet and Cardiovascular Disease	Committee on Medical Aspects of Food and Health Policy (COMA), 1984
Eating for a Healthier Heart	British Nutrition Foundation and the Health Education Council, 1985

Certain common guidelines have come from these reports:

- We should eat more vegetables (including potatoes), cereals (including bread) and fruit
- We should eat less sugary, fatty and salty foods and drink less alcohol
- We should eat a wide variety of foods
- Expectant and nursing mothers may need additional vitamins

- We should not become overweight
- We should exercise moderately and frequently.

The guidelines proposed by the reports, particularly the NACNE report, have been welcomed by many people, but others have criticized them as being unrealistic. Studies have shown that it is possible to achieve the goals of the reports in changing one's diet. Dietary advice, however, should be given so that people do not over-react and so that they adapt over a period of time.

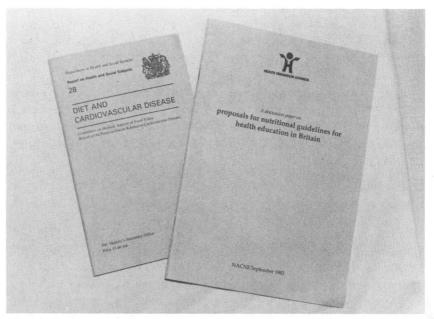

The NACNE and COMA reports

The NACNE report

The National Advisory Committee on Nutrition Education (NACNE) was formed in 1979. NACNE is made up of representatives from medicine, nutrition education, the Ministry of Agriculture, Food and Fisheries, the British Nutrition Foundation, DHSS (now the DoH), the Scottish Health Education Group and the academic community. The NACNE report, *Proposals for Nutritional Guidelines for Health Education in Britain*, was a consensus of these representatives' opinions based on considerable expertise and experience. Although many aspects of the report were not new, it had a mixed reception, probably because it challenged a number of deeply entrenched ideas on nutrition.

The report's rationale was: the British diet is a serious threat to the life and health of the nation and must be changed and modified through a variety of policies. Similar conclusions have been made in other countries, particularly in the United States.

A major health problem in Britain for some years has been obesity. It is estimated that 40 per cent of the population is overweight. NACNE reported that in certain individuals even being slightly overweight constituted a health risk. Degenerative changes could start in people even in their teens, resulting in a heart attack in perhaps 20 years' time.

Diseases of poverty associated with low nutritional status have virtually been eliminated in this country. The concept of a 'balanced diet' consisting of the recommended levels of protein, carbohydrates, vitamins and minerals is an anachronism. The balanced diet should now be made up of foods which are known to have a lower health risk. Evidence used by NACNE suggested that intakes of fats and sugar should be reduced, but the consumption should be increased of potatoes, bread, fruit and vegetables.

The question of fat intake

An overall reduction in the population's fat intake is needed. In the average British diet energy supplied in the form of fat is about 42 per cent of the total energy. NACNE recommended that this should be reduced to 30 per cent and that only a third of this should be saturated fat. This entails the encouragement of selective breeding of lean animals for meat, but even lean meat can contain 25 per cent fat. However, fortunately more of this is of the unsaturated type.

The old idea of fat surrounding the Sunday joint to give flavour must be abandoned. It is questionable whether it really contributed to flavour anyway. The fat content of most products must be reduced, as often fat is used as a cheap filler.

Fat is used in the manufacture of biscuits, cakes and flour confectionery and similarly must be reduced. New special starches can give the products some of the properties associated with fats, so fat levels can be reduced. The NACNE report also emphasized that food products must be clearly labelled as to their content, particularly of fats (see section on Nutritional Labelling, page 30).

The COMA report took the fat issue a stage further. This will be discussed later (see page 25).

The question of carbohydrate intake

For people on diets to lose weight, reduction of carbohydrate in the form of potatoes or bread has always been recommended. The poor reputation of carbohydrates was encouraged by the popular diets which were based on a low calorie intake. Some foods contain carbohydrates, particularly sugar, which are readily and quickly absorbed by the body (the energy from these can be called *simple calories*). Foods such as potatoes and wholemeal bread contain complex carbohydrates which are broken down slowly in the body to release calories gradually (*complex calories*). These foods are generally starchy and digest more slowly, thereby releasing their calories to the body over a longer period and imposing less strain on its metabolism.

NACNE recommended that starchy carbohydrates should supply 50 per cent of the body's energy requirements. For most of us this would represent increasing our intake by about a quarter of bread, potatoes, fruit and vegetables.

The question of fibre

Many of the starch foods recommended by NACNE are rich in fibre. In fact, the report proposes that people should eat 30 g of fibre a day. Since only 5 per cent of the population eats this amount, it constitutes a significant increase for most of us with a daily average of 30 g.

Fibre can be defined as any plant substance that is not digested by human enzymes. It is best obtained by eating foods containing more fibre and not by taking dietary supplements of crude fibre. Some fibre, for example from cereals, may combine with minerals and therefore reduce their intake into the body. This has been found to happen with calcium and high fibre diets. Plant products rich in **cellulose**, **hemicellulose**, **pectin**, various **gums** and **protein** material are good sources of fibre.

Fibre-rich foods produce greater bulk in the stomach. Digestion therefore takes place more slowly, releasing energy gradually to the body. Some fibre, for example from cereals, absorbs water and swells, whereas other fibre, for example from fruit, actually encourages the increase in bacterial numbers which gives greater bulk. The intestines work efficiently with this bulk and constipation is eliminated.

Insoluble fibres such as cellulose, wheat fibre and fibre from fruit or vegetables, show their greatest effect on the operation of the intestinal tract. Some *soluble fibres* absorb water and form gels which reduce the absorption of undesirable substances such as cholesterol. Recently apples were cited as being good in reducing cholesterol because of their pectin content. There is some truth then in the old saying 'an apple a day keeps the doctor away'.

Foods low in fibre are linked to cancer of the bowel. There are currently, three theoretical mechanisms as to how this might happen:

(a) Faeces which are small and hard have a higher concentration of substances which may be carcinogenic. Greater fibre intake dilutes this concentration of carcinogenic material, thus reducing the risk of cancer

(b) High fibre diets speed up the passage of food through the body thus reducing contact time between these disease forming substances and the gut

(c) High calorie intake has been associated with a higher intake of carcinogenic substances. High fibre diets actually reduce calorie intake, hence their use in 'Slimming foods', and thereby reduce the intake of carcinogens

The complete picture on fibre is still not known but it would appear that a gradual increase in dietary fibre would be beneficial to health. There are problems, however, as sufficiently high doses are unpalatable to some individuals. Problems associated with flatulence, nausea, and diarrhoea have been reported. This desire for more fibre has caused the public to turn against highly refined foods and buy wholemeal products and so-called 'health foods'.

The question of salt intake

The NACNE report recommended a reduction in salt intake to a half or even a quarter of the present 12 g per day. This recommendation was based on an apparent link between high salt (ie sodium) intakes and higher resultant blood pressure causing hypertension. Low occurrences of hypertension were found in populations with a high dietary intake of potassium and low intake of sodium. Similarly, blood pressure was found to fall in some individuals on low sodium diets.

A substantial amount of salt is contained within processed foods and, as such, these provide us with more than enough salt for normal uses. Some people take many times the amount of salt the body actually needs. However, since the NACNE report was published, research has shown that only certain individuals are susceptible to high sodium levels. In some individuals, the reduction of salt levels may even be harmful. Which of us is susceptible to high sodium levels? Most of us will not know the answer, so perhaps a reduction in salt may be a wise precautionary move.

The question of alcohol

Alcohol is not part of a normal diet, but for some people it contributes a considerable intake of calories, estimated at between 4 and 9 per cent of their total energy intake. The effects of excessive alcohol intake are well known, but other serious disorders can develop:

- Alcoholics frequently suffer from malnutrition, and particularly from vitamin deficiencies
- Cirrhosis of the liver can be a result of high alcohol consumption
- There is evidence of depressive effects and even brain damage

However, small to moderate alcohol consumption is believed to be generally beneficial. This level should not exceed 4 per cent of energy intake, slightly less than one pint of beer per day. There is evidence that alcohol in these amounts actually causes a reduction of cholesterol in the blood, which reduces risk of heart disease.

Summary of NACNE proposals

The long term proposals of NACNE are summarized below:

1 A standard approach to dietary recommendations is required
2 The definition of ideal energy intakes should be related to the maintenance of ideal body weight, with moderate exercise
3 The dangers associated with being overweight should not be exaggerated compared with the risk of smoking
4 Fat intake should be restricted to 30 per cent of the total energy intake
5 Saturated fatty acids should not exceed one-third of the above fat level, ie no more than 10 per cent of the total energy intake

6 No recommendations were made to increase polyunsaturated fatty acids (see COMA on pages 25–28)

7 No recommendations to lower cholesterol

8 Sucrose intake should be reduced to an average intake per person of 20 kg per year

9 Fibre should be increased to 30 g per person per day

10 Salt intake should be reduced by 3 g per day

11 Alcohol intake should be restricted to 4 per cent of the total energy intake

12 Protein levels should remain the same, but more vegetable protein should be consumed

13 Mineral and vitamin intakes should match the recommended dietary allowances (RDA) listed by the DHSS (now the DoH)

14 Better labelling of processed foods should be encouraged

The COMA report

The NACNE report was followed fairly quickly by the report produced by the Committee on Medical Aspects of Food Policy on *Diet and Cardiovascular Disease*, generally known as the COMA report. This report was concerned mainly with 'the British epidemic of the 80s' – cardiovascular disease.

The recommendations of the report were developed after studying over 600 scientific papers. These recommendations are given below.

(a) For producers, manufacturers and distributors of food

1 Fats, **oils** and products containing at least 10 per cent fat should be clearly labelled to give the percentage by weight of fat, and the percentage of saturated, polyunsaturated and trans-fatty acids*. Simple labelling codes should be encouraged

2 Drinks containing more than 1.2 per cent alcohol should have their alcoholic content printed on the label

3 Foods with lower contents of saturated fatty acids, trans-fatty acids and salt should be made available

*'trans' refers to the structure of the fatty acid; this type may be harmful and accumulate in the body. The opposite is 'cis', which is more easily metabolised

(b) For the general public

1 Fat and saturated fat intake should be reduced
2 Simple sugar and salt intakes should not be increased
3 Excess alcohol should be avoided
4 Fibre-rich carbohydrates should be increased in the diet
5 Obesity and smoking should be avoided

(c) For medical practitioners

Those at risk should be identified and advised as to their diet

(d) For the government

Means should be found to educate the population in eating habits and exercise which will minimize the risk of heart disease and obesity

There are areas of interest common with the NACNE report but again absolute proof of dietary links to heart disease cannot be given.

Unsaturated fats

Surprisingly in the COMA report there were no recommendations for change in the consumption of polyunsaturated and mono-unsaturated fatty acids. There is evidence that if the ratio of polyunsaturated to saturated fatty acids (P/S) is increased to about 0.45 there are definite health advantages. On average in Britain the P/S ratio is only about 0.28. There are two main advantages of increasing polyunsaturated fatty acids. Firstly, two of these acids, linoleic and linolenic, are essential to life, and therefore should be called 'essential fatty acids'. Deficiency of these acids has been found in cases of multiple-sclerosis (deficiency in the mother), skin complaints and abnormal growth. A normal person should have an intake of essential fatty acids which corresponds to about 1 per cent of the person's normal calorie requirement.

Secondly, although most polyunsaturated fatty acids (PUFAs) are not essential, they play an important role in reducing the risk of heart disease. Many people are aware of this and the sales of soft-spread margarines richer in PUFAs confirm this. PUFAs play saturated fatty acid (skeleton)

$$C-C-C-C-C-COOH$$

unsaturated fatty acid

$$-C=C-C=C-COOH$$

trans-fatty acid

$$C-C-C-COOH$$
$$\|$$
$$-C-C-C$$

cis-fatty acid

$$HOOC-C-C-C$$
$$\|$$
$$-C-C-C=C-C$$

Saturated and unsaturated fatty acids

a vital role in helping reduce cholesterol in the blood.

Research has shown that cholesterol will combine with lipo-proteins (made of a fat-type substance joined to protein) of varying densities. High density lip-proteins combine efficiently with cholesterol and thus reduce the risk of heart disease. Conversely, low density lipo-proteins are inefficient in combining with cholesterol which can then be available to cause problems in the blood vessels. Some people are fortunate in having high density lipoprotein-cholesterol, whereas others, at greater risk, have low density lipoprotein-cholesterol. Work has shown that by increasing the level of PUFAs in the diet the former case of high density lipoprotein-cholesterol is encouraged, thus reducing heart disease. A small to moderate intake of alcohol may also show a similar beneficial effect.

Recent work has shown that in some Mediterranean countries, for example Italy, there is a low occurrence of heart disease but a relatively high fat intake. This seemed to go against everything recommended to date. It was found that the widespread use of olive oil and similar products gave a substantial intake of mono-unsaturated fatty acids, for example oleic acid. The results of this diet were shown in a reduction of blood cholesterol comparable to diets low in fat.

Diets rich in fish, particularly fatty fish, have been shown to have a protective effect against strokes, heart disease and diabetes. The diets of Greenland Eskimos and of those in Japan have illustrated this. Fish oils are rich in very unsaturated *Omega 3** (also known as n–3) fatty acids, whereas vegetable oils have more *Omega 6** (or n–6) which are not so beneficial. There are seven Omega 3 (n–3) fatty acids in fish oils, of these two dominate: *EPA* (eicosapentaenoic acid) and *DHA* (docosahexaenoic acid). We should eat more fish to increase our intake of Omega 3 (or n–3) fatty acids.

The role of calcium in the diet

Interest has been shown recently in the role of **calcium** in the diet and the benefits of a higher intake to the health of individuals. Dairy products are the main supplies of readily available calcium in the diet. The calcium in milk, yoghurt and cheese can be readily absorbed, constituting nearly 60 per cent of our calcium requirements.

The absorption of calcium is influenced by vitamin D obtained from oily fish, margarine and breakfast cereals, or by the action of sunlight on the skin. Lack of sunlight for some people, particularly dark-skinned people living in Britain, can lead to vitamin D shortage and poor calcium absorption which ultimately leads to the disease **rickets**. Cereals supply some calcium but sometimes this is made unavailable as it can combine with **phytic acid** a substance found in wholemeal products.

During rapid growth in adolescence, additional calcium is required for bone growth. Similarly, expectant and nursing mothers need additional calcium. Particular attention has been drawn recently to the problem of calcium loss from the bones during middle-age, particularly amongst women. Calcium is lost from about the mid-thirties in some people, but post-menopaused women lose more because of hormonal changes and often lack of exercise. The condition of *osteoporosis* can develop if the loss of calcium continues. The spaces within the bones increase so making them weak and fragile leading to more fractures. A

---

*For chemists: this indicates the first double bond or point of unsaturation in the fatty acid.

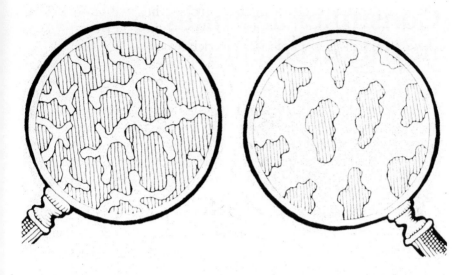

The difference between an osteoporotic (left) and a normal (right) bone

generous calcium intake slows down this loss, thus delaying the onset of osteoporosis.

Calcium has also been found to have other functions. It appears to be an activator of some enzymes and hormones, such as insulin which controls sugar metabolism in the blood. There is also a hint that a generous calcium intake will reduce coronary heart disease. In some patients suffering from high blood pressure a calcium deficiency has been discovered but these aspects are difficult to prove conclusively.

Further Reading

National Dairy Council, *Our Daily Calcium* (NDC, 5–7 John Princes Street, London, W1N 0AP
National Dairy Council, *Calcium and Health* (more advanced!) (NDC – see above)

Consumer attitudes to nutrition labelling

The reports discussed in this review indicate the benefits of, and the public's desire for better labelling, particularly concerning nutrition. It is a legal requirement for certain foods to have nutritional labelling; for some foodstuffs it is voluntary, but gradually the requirement will be extended to all foods. Research has shown that people want and welcome nutrition information on labels. But do they know what it means?

Most people have heard of *fat, proteins, sugars, calories, carbohydrates, salt* and *vitamins*. Fewer people know of *kilocalories* and fewer still know of *joules* or *kilojoules*. Terms such as *polyunsaturated, saturated, fatty acid* and particularly *trans-fatty acids* are rarely understood.

Simple diagrams, bar charts and 'traffic lights' are the most popular means of conveying nutritional data. Complex lists of nutrients, given as mg/100 g, are rarely understood. If consumers are to control their daily intake of nutrients, they must be able to estimate quickly the quantity of a particular nutrient in a food. Just imagine how long it would take to do a week's supermarket shopping if every label had to be read and nutrients assessed in each product!

NUTRITION

Instant mashed potato is fortified with Vitamin C which we need daily to ensure good general health.

A SERVING = ¼ OF THE SACHET - MADE UP ACCORDING TO INSTRUCTIONS

AVERAGE COMPOSITION	PER 140g (5oz) serving	PER 100g (3½ oz)
Energy	365kJ/87kcal	261kJ/62kcal
Fat	1.0g	0.7g
Protein	2.7g	1.9g
Available Carbohydrate	17.8g	12.7g
Fibre	2.0g	1.4g
Added Salt	1.1g	0.8g
MINERALS/ VITAMINS	% RECOMMENDED DAILY AMOUNT	
Vitamin C	93%	20.0mg

THIS PACK CONTAINS 8 SERVINGS

INFORMATION

Food label showing nutritional information

Conclusion

The diet of most people is gradually improving; people are becoming increasingly aware of the trends in nutrition and the benefits to health these can bring. Unfortunately many aspects of nutrition cannot be proved beyond all doubt. We are all different and react to nutrients in different ways, so rigid rules should not apply. Guidance on nutrition in a clear, understandable way is what most people appreciate.

Further Reading

National Dairy Council, *Off to a fresh start with food and fitness* (National Dairy Council, 5–7 John Princes Street, London, W1N 0AP)

3 A look at modern food processing

In Britain around three-quarters of our food has been processed to some degree or other. A highly competitive modern industry has developed to produce wholesome, safe and nutritious foods. In fact, without a food industry of this type, our modern way of life would be impossible – perhaps some of us would prefer that! The Industry has to deal with raw materials which are becoming more expensive; with a vast amount of controlling legislation; with customers whose loyalty has to be earned; and with constant inspection and control.

The Food Manufacturing Industry transforms farm produce after harvesting into a range of edible food products. Simple processing can occur on the farm or soon afterwards, for example, butter-making, milling, freezing vegetables, and curing meat. More complex processing involves the blending together of ingredients in the right proportion, often with the aid of addi-

The main packing hall at a McVities factory

tives, to produce new and varied food products. These products usually take on a form which is convenient to use. The growth of convenience foods has been quite dramatic and is continuing, even though many people buy some so-called 'health foods'.

Convenience foods can be defined (National Food Survey Committee) as foods for which the degree of culinary preparation has been carried to an advanced stage by the manufacturer, and which may be used as labour-saving alternatives to less highly-processed products. This preparation by the manufacturer will include cleaning, sorting, and grading (quality assessment) the raw materials. Some form of size reduction, for example slicing, may be required. Other ingredients may need to be added by mixing. Liquid products will need to be filtered to remove suspended matter and often will require concentration. (For details of processes see: R. K. Proudlove, *The Science and Technology of Foods* (Forbes, 1985).) Although advances have been made and are continually being made in these preparation processes, the most significant advances in recent years have been made in *preserving* the products and in their *packaging*.

Methods of food preservation, and production

The traditional methods of food preservation such as **canning**, **drying**, **curing**, **freezing** and **smoking** continue to be used to produce an enormous range of food products. Although some of these processes have declined, freezing of foods is still expanding. The **chilling** of food has shown enormous growth in the last few years and has made in-roads into the traditional frozen food market. Although **microwave** ovens have been used in the home for some time, the same principle of operation is now being applied to commercial operations. The growth of 'snack foods' has been rapid and many of these foods have been produced by the process of **extrusion** to give a great range of product shapes, sizes and texture. The most modern method of food preservation, **irradiation**, has been thoroughly investigated for over thirty years, but at the time of writing is still not legally permitted in the UK.

Once preserved, foods must be protected from recontamination by micro-organisms and from the 'elements'. Increasingly sophisticated packaging materials and techniques are available to the food processor. These packaging materials enable the product to be decorated attractively and to carry all the necessary and legally required information on the label.

Below, these modern methods of preserving and packaging foods will be reviewed. A brief look will also be given at some traditional methods and how improvements have been made to them.

Canning

Canning can be regarded as the first large scale food processing method. It was invented by Appert and was established during the last century. Initial attempts were erratic but some samples of canned foods survived well into this century. Canning really became popular when the double-seam method of hermetically sealing the can became established. This facilitated rapid filling of cans which were efficiently sealed and sterilized. However, the tinplate used to make cans has become increasingly expensive – in some cases the can could cost more than the contents. Thinner tinplate has therefore been developed and reinforcement rings or undulations have been used to add strength to the can to reduce the risk of denting.

Pressure from frozen foods, in particular, and a greater demand for fresh produce has caused a decrease in canned food production. Quality is vitally important but it must be achieved using a high volume of low value products. This is often difficult to attain and has to be accomplished within a number of constraints. Weather affects the sale of canned foods quite significantly. Adverse conditions can cause a poor harvest of fresh produce so canned products (from the previous season) are bought. However, the quality of raw materials for canning will also be poorer if the harvest is poor. Consequently, the following year canned products may be noticeably inferior, particularly if the fresh market is good. Fortunately, there will always be some products that must be canned. Could you imagine frozen baked-beans in tomato sauce?

Modern canneries operate through contract growing of their raw materials. The farmer is provided with the seed, fertiliser, pesticides and back-up technical support. He grows the crop and is paid according to yield and quality. In this way the factory will know exactly how much raw material is available, at what time and of the right quality.

Peas are still the most important seasonal product. As the day of harvesting approaches, regular tests are performed on the peas so that they are picked with exactly the right degree of tenderness and consequently the right sweetness. A special machine called a *tenderometer* is used which crushes a certain volume of peas and the force required is recorded. This gives a measure of tenderness.

Peas are harvested by viners, which can also remove them from the pods. The peas start to deteriorate fairly quickly, with sugar being converted to starch. The viners are therefore in radio contact with transport vehicles and the factory to ensure that the peas are canned within the shortest possible time, usually 1.5 hours.

On arrival at the factory the peas are weighed and are then subjected to a number of cleaning operations:

(a) Stones and unwanted dirt are removed on vibrating conveyors

(b) The peas are passed to the blancher where they are heated in near-boiling water for a short time to inactivate any enzymes present and to drive out trapped air. The **blanching** water also cleans the peas

(c) The peas are immediately cooled after blanching, then inspected on moving belts

(d) The right quantity of peas is automatically dropped into each can and brine is added. This process is often under electronic control and the brine is automatically heated on its way to the filling line

(e) The cans are then sealed, at rates that can exceed 1200 cans per minute

(f) The cans are sometimes washed and cooled before passing to the retort or cooler. Many factories employ continuous or **hydrostatic retorts** which allow continuous production unlike the old batch-type retorts

(g) After retorting, high-speed labelling and wrapping of cans takes place. Shrink-wrapping of finished cans is often employed

Many products are canned in this way. Surprisingly there have been a number of innovations in canned foods of this type. Fruit

Canning line

is often canned in juice now and not in thick syrup as it has been
for many years. However, the juice is probably from a different
fruit! Colours in processed products, for example processed peas,
are now viewed with suspicion and salt levels in brines have been
reduced. Canning, because of the heat process, alters products,
particularly their colour and flavour. Tomato soup is a typical
example of an artefact of canning in its flavour and colour; a
natural tomato flavour and colour in soup is unacceptable to most
people.

Considerable development work has been undertaken on the
traditional can:

- The can became lighter and a one-piece construction, free of
  seams, has been developed
- Aluminium has become very popular and now accounts for 50
  per cent of the beverage can market
- Glass in its new lightweight but strong forms has made some
  in-roads into the can's monopoly, particularly in delicatessen
  products
- Soups are now being sold in cartons rather than cans
- In many countries, particularly Japan, flexible pouches have
  almost completely replaced the metal can

The next leap forward will be the perfecting of a plastic sealable can. Development work is being undertaken to produce such a can which is suitable for a great range of products. Problems which have to be solved include seaming difficulties and discoloration of the can by the product it contains.

Drying

Drying is one of the oldest methods of food preservation. The traditional method of sun-drying is often long and obviously uncontrollable. Nevertheless it is still practised in many parts of the world. Food dehydration is an important operation in the Food Industry in producing products for direct consumption or a vast array of ingredients for adding to other foodstuffs.

Drying is carried out on a food for a number of reasons:

- The reduction of bulk by removing water lowers transport and handling costs. (This is why milk powder is sent to famine areas rather than liquid milk)

- By reducing the moisture level to below a critical level, perhaps around 2 per cent, bacterial and enzymic action is reduced or prevented – hence the preservative action of dehydration

It is, however, impossible to remove all water as some is tightly bound to various food components such as some proteins.

In selecting modern dehydration plant manufacturers have to consider a number of variables. A study must be made of the drying characteristics of the food being processed:

- Some foods dehydrate easily to produce a dry product which readily rehydrates in water when used. This is a vital characteristic if the product is to be used as a convenience food

- Some products contain soluble materials which move to the product's surface during drying and, together with proteins present, form a skin on the surface, known as **case hardening**. Case hardened products do not always dehydrate completely and, because of the skin present, equally do not reabsorb water readily. One trick tried with dried peas was to prick each pea so that the water could escape during dehydration through the hole produced and could equally rehydrate through the same hole

- Some products shrink readily during the drying operation, to produce dense products which do not rehydrate easily. Generally, slow drying, such as sun-drying, causes the maximum shrinkage in a production. Rapid drying methods produce less shrinkage and a product which has a lighter, porous texture, like a stock cube. Rehydration of this type of product is much more rapid and complete

In modern processing many liquid products are dried. One of the most suitable methods for drying liquid is **spray drying**. Many liquid foods are heat-sensitive and are readily damaged by excessive heating. Spray drying avoids heat damage by spraying or atomising the product in a chamber where the liquid droplets meet a blast of warm air. Drying of the droplets is rapid and there is little or no heat damage. Many products can be produced by this method from milk to chicken powder, where the chicken flesh is converted to a slurry before spraying.

Although single-stage spray drying is used extensively, it tends to produce a fine, dusty product like flour. Dried milk produced by this method is difficult to wet and floats on the top of water, but when stirred often forms lumps. Modern processes are in two stages and involve the process of **fluidized bed drying**:

1 The first stage allows the retention of a lot more of the water from the milk. In this way, the fine particles, formed as the milk droplets dry, tend to stick together

2 The second stage involves a fluidized bed drying system where warm air is blown through the particles to complete the drying. Air blown through a powder tends to give the powder the properties of a liquid: for example, the ability to flow, hence the name 'fluidized bed'. The particles which have stuck together form granules which are like sponges readily dispersing and dissolving in water

The concept of fluidized bed drying can be applied to the drying of other products such as peas or dried vegetables. It is possible to make the process continuous and the dried foods can be packed into moisture-proof pouches at the end of the drying tunnel. This method causes little shrinkage of the product as it is fairly rapid and can only be carried out on food particles not much larger than a pea.

The best quality dehydrated products are produced by the process of **freeze drying**. The process initially involves freezing the product which is then subjected to a high vacuum and sometimes to a very small amount of heat (accelerated freezing drying). As a result of this treatment the ice in the product does not melt but goes directly to water vapour by the process of sublimation. This leaves a product which is honeycombed and which readily reabsorbs water.

The advantages of freeze drying compared with other drying methods are considerable:

- The original colour, texture and most of the flavour of the product is retained
- Vitamins and other nutrients suffer less damage
- The product has the best rehydration properties of all dried products, together with a longer shelf-life

The main disadvantage of the process is that it is slower and therefore, more expensive.

The difference between a freeze dried product and a conventionally dried product can be seen in instant coffees. Freeze dried coffee has retained a greater amount of aroma and flavour compared with the conventional spray dried coffee. Why not use frozen instead of dried products? For many people, dried products are not so convenient as they require rehydration, but refrigeration is not required to keep them for long periods under varying temperature conditions. A number of products, like coffee, are produced for their flavour or aroma, for example, mushrooms, asparagus and sea-foods. Freeze drying is an excellent method of preserving these because of its ability to retain flavour.

Dried products can be spoiled by micro-organisms, particularly fungi, if their moisture contents rise above a certain figure. This figure may be quite low so effective packaging, ideally a re-closeable tin, is essential. It is possible to make water unavailable to micro-organisms, and this is exemplified in the large use of preserves, sugar confectionery and salted products. Substances which absorb or readily dissolve in water can make the water unavailable to micro-organisms, by lowering what is known as the **water activity** (a_w) of the product. Pure water has a water activity of 1.0, foods have a lower figure than this but may have a figure of 0.9 or above. Most bacteria can grow at a_w greater than 0.8, and some **yeasts** at 0.6. Drying lowers the a_w to well below this figure and in jams, preserves and salted products the a_w is lowered by the osmotic effect of the sugar or salt. Intermediate moisture foods have been developed in which the moisture level is reduced somewhat and then the remaining water is made unavailable to micro-organisms by the action of a substance such as **glycerol** (glycerine), with the help of salt or sugar. Unfortunately these foods proved to be rather unpalatable and have been confined to the pet-food market.

Chilling

In 1986 this relatively new sector of the Food Industry was estimated to be worth up to £10 billion in sales. The growth of chilled foods has been remarkable, but is undoubtedly due to efficient transportation and rigid quality control. The major advantage of chilled foods is that they are considered 'fresh'. Major retailers have pioneered the growth of these foods, particularly as 'quality foods' can be chilled and justify the rigid controls necessary.

What are chilled foods? **Chilled** foods are not frozen and must never be frozen to justify the title 'chilled'. They are perishable, but are kept wholesome for a given time by being maintained at a temperature of between $-1°C$ and $8°C$ (*Guidelines for the handling of chilled foods*, Institute of Food Science and Technology, 1982). Of course, great care must be taken in determining the appropriate storage temperature (recently listeria has been found in certain chilled products). Ideally a temperature of less than $4°C$ should be used. Chilling thus extends the shelf-life of a fresh food (a) by slowing down the changes within the food which cause deterioration (usually due to enzyme action) and (b) by retarding the multiplication of micro-organisms.

The basic aim behind the chilling process is to ensure that food poisoning organisms do not grow. The optimum temperature at which a micro-organism grows varies with the type of organism (see Chapter 4) and fortunately it has been found that food poisoning organisms grow more readily at relatively high temperatures. Therefore the temperature range chosen for chilled foods is one in which some organisms (but not those causing food poisoning) can grow. These will eventually spoil the product, but there should be little risk of food poisoning because the temperature has not been high enough to cause the multiplication of the food poisoning organisms.

Examples of chilled foods are numerous including dairy products, meat, fish, salads, fruit and vegetables. However, a major growth area is that of convenience or prepared complete meals

Range of supermarket chilled foods

such as pizzas, pies and pasta products. There has also been a significant growth of fresh snacks and sandwiches, the latter having seen off the infamous British Rail sandwich in some regions.

Innovative products are able to get into the market remarkably quickly. Unlike frozen foods where storage facilities are essential, with chilled foods it is the *distribution system* which is all important. This distribution system is usually under computer control and is a tribute to modern planning within the Food Industry. The system works basically as follows:

- The stores within the retailing group have computer links to their head-office and pass their daily requirements through the links so that the head-office can then request the food manufacturers to produce the right quantity of each product

- The chilled products are transported to a central distribution point, where the deliveries are broken down into small lots for delivery to individual stores

- The central distribution point receives computer data from the head-office as to where to deliver the chilled foods and in what quantity

- Experience has shown that it may take 4 hours, for example, to reach a particular store, so the transport vehicle leaves with its load of chilled foods at about 4.30 am to arrive at the store just before opening, usually each day

- The product will probably only have a shelf-life of a day or two and so fresh deliveries are needed continually

The product mix going to a particular store could be enormous and so computer control is essential.

The storage life of chilled foods is dependent on the maintenance of chilled conditions at all times and on the hygiene of processing and handling. During handling and production, products are kept in chill stores and then loaded directly into chilled vehicles. The vehicles either have their own refrigeration systems operated by compressors worked by diesel engines, or they are chilled by liquid gases, usually liquid nitrogen. This latter system requires little maintenance but can only maintain the chilled food at the temperature set and cannot be used to reduce temperature.

The life of chilled foods can be extended quite considerably if they are stored in the presence of certain gases, particularly carbon dioxide. Similarly the level of oxygen may be reduced or nitrogen added to effect the same result. The gases slow down microbial deterioration of the product but allow respiration of living foods, such as fruits, to continue, but at a retarded rate. This method of storage is known as **controlled atmosphere (CA) storage**.

A method of reducing the atmospheric pressure in a store, **hypobaric storage**, has been developed but found to be too expensive in operation. In this method reducing the atmospheric pressure by applying a vacuum causes a reduction in oxygen levels and retards the spoilage of many products and the ripening of fruit.

Some of these ideas have been extended to use within packaging materials. Some packs have been flushed with gases, for example smoked fish with nitrogen, which causes an increase in shelf-life. Carbon dioxide has also been tried for this purpose. This type of packaging is known as **modified atmosphere (MA)** packaging.

A recent development, which is simple and effective, is to coat some fruits and vegetables in an edible substance which slows down their respiration and subsequent deterioration. The substance known as '*Pro-long*' is a mixture of cellulose and wetting agents. The product is dipped into a 'Pro-long' solution and allowed to dry. The 'Pro-long' coats the fruit and controls its respiration by allowing oxygen to enter but little carbon dioxide to escape. By this means the respiration of the fruit slows down and, as a consequence, it takes longer to ripen and deteriorate. Unfortunately, each fruit needs its own concentration of 'Pro-long', but many fruits have experienced a doubling of their shelf-life when the system is used with chilling.

Freezing

Freezing preserves foods by two means:

(a) the obvious low temperature inhibits micro-organism growth

(b) the formation of ice during freezing effectively removes water from the food and shows the same preservative effect as dehydration

The rate at which a food is frozen is important. Slow freezing leads to the formation of large ice crystals which draw water from the cells of the food, causing irreversible damage to the product. Rapid freezing leads to small ice crystal formation with less cellular damage. However, what is *rapid freezing*? A pea can be frozen in a few seconds, a side of beef may take 36 hours. It has become apparent, recently, that it is only possible to freeze small food items such as peas rapidly; all other products must be frozen at a slower rate than sometimes equates with maximum quality in the finished product.

After freezing, frozen products are stored at $-18°C$ or less (this temperature being the EC directive). Storing at $-25°C$ or even $-29°C$ has shown distinct advantages in large scale or long term storage. Fluctuating temperatures during storage can cause a number of problems; storage at $-25°C$ or less minimizes these.

- Fluctuating temperatures encourage rancidity in products containing fat, such as minced beef

- Green vegetables start to lose their green colour, **chlorophyll**, and eventually change to a dull greyish brown colour due to the change of chlorophyll to **pheophytin**

- A serious problem can occur due to varying temperatures when ice crystals undergo sublimation to form water vapour. This loss of water vapour causes weight loss in the product and icing of evaporation plates in the cold store or the accumulation of 'frost' within the product's packaging. Small ice crystals are less

Range of frozen products

stable than larger ones and are more readily lost by sublimation. However, some water vapour will condense and refreeze on the larger ice crystals, thus making them larger. A product containing large ice crystals may produce quite a pool of 'drip' around it during thawing. This is shown in badly frozen fish

There are three large sectors in the frozen food market taking up 75 per cent of all sales, these are *meat* products, *ice-cream* and *fish* products. *Potato* products, particularly frozen chips, have about 12 per cent of the market, with *frozen desserts* having about 7 per cent. There is a growth on frozen convenience products (although all are convenient!) particularly in complete meals. These are often developed with the microwave oven in mind. Chilled foods have made an obvious inroad into the frozen food market which would have expanded considerably more without this competition.

As there can be no waste from a frozen product, very careful preparation of the food raw materials is essential. Cleaning, sorting and grading operations have to be very thorough. All vegetables for freezing have to be blanched before freezing. This is usually carried out by heating in boiling water or steam for a few minutes. If blanching is not carried out, naturally-occurring enzymes in the food will produce flavour, colour and texture changes during frozen storage. Some products are stored for several months before reaching the retailer and so shelf-life has to be much longer than domestically frozen produce. In the busy harvesting season there is now a tendency for products to be frozen and stored in bulk. This frozen bulk product is subsequently broken down and repacked into retail sized containers nearer to the date required. A large service industry has thus developed to supply cold-storage and repacking facilities for the frozen food industry.

There are three main methods of freezing. The oldest method is by **plate freezing** where the product is in contact with a refrigerated plate and produces the traditional flat packs of frozen produce. This method is much less common but is used mainly for producing fish blocks.

There are a number of variations of the **blast freezing** method which blows cold air at a product of any shape or size. Variations of this include **fluidized bed freezing** which corresponds to the fluidized bed drying in its mode of operation. Here products can be **individually quick frozen (IQF)** and then packed in the familiar polythene bags.

The third method is **cryogenic freezing** which employs a very cold liquefied gas as its freezing agent. The most popular method uses liquid nitrogen which boils at $-196°C$. The product, on a conveyor, passes under sprays of liquid nitrogen which converts to the gas and may be lost to the atmosphere. The method allows very quick freezing of small products such as raspberries or prawns, but is too slow for large items as the freezing effect of the liquid nitrogen has poor penetration ability. In Europe, liquid carbon dioxide at $-78°C$ is preferred to nitrogen.

Like chilled foods, frozen products have to be maintained at the required temperature during storage and transportation. This is obviously more difficult to carry out than with the higher temperature chilled foods. During the 'cold-chain', from the freezing plant to the domestic user, there is considerable fluctuation in temperature and a general raising of the product temperature. It is a tribute to modern frozen foods that the quality is often very high and rarely unacceptable.

Extrusion cooking

There is now a vast array of products produced by **extrusion** including snack foods, textured proteins, breakfast cereals, pet and baby foods. The technology responsible for this has only become widely used in the last few years, although it dates back to 1935 when macaroni was first extruded. An extruder will form, texturise and cook a food within one piece of equipment. The cooking aspect of extrusion is at a **high temperature for a short time (HTST)**. Bacteria, enzymes and heat-sensitive toxins are destroyed quickly without unnecessary heat damage occurring in the product.

An extruder consists of a large screw, rotating within a barrel that becomes more tightly fitting along the length of the screw and is also heated. The barrel is effectively divided into three sections and the screw's thread varies to meet the requirements of each section:

- In the first section the food ingredients become fully mixed and become compressed

- In the second section, the compression section, the food is worked and compressed, causing its temperature to rise rapidly

- The third section is the metering section where uniform temperature is achieved and the product is metered out through a specially shaped die

Extruded snacks during manufacture

More recently twin screw extruders have become more popular as they are more economical to use and are more controllable. Two co-rotating screws are used in most applications.

The raw materials used in extrusion are mixtures of *cereal fractions, starch, proteins, fat* and a number of *additives* such as *emulsifiers*. Usually water is added to give the raw materials a moisture content of about 25 per cent. This mix is preheated often by steam injection. On entering the extruder the raw materials' temperature rises rapidly due to friction, heat transfer from the barrel and sometimes steam injection into the barrel. Temperatures may reach 150–200°C, but usually for only a fraction of a second. Heat damage to the product is thus avoided. The combination of heat and pressure in the compression section of the extruder cooks the raw materials and mixes them into a homogeneous mass. Many changes occur in the materials, starches become gelatinized, flavours develop, proteins become denatured and some products are broken down into simpler substances whereas others are made into large polymers.

After passing through the metering section, the heated mass is suddenly released through a shaped die. There is a sudden release of pressure which causes the product to foam or puff internally as water is converted into steam. Moisture is lost rapidly at this stage and the product cools quickly. A rotating knife cuts the extruded product into short lengths as it comes through the die.

Most extruded products are somewhat fragile, with a puffed or cellular structure. They can be covered with flavours, colours, salt or sugar and are usually ready-to-eat. Their texture and subsequent mouthfeel, will depend upon the nature of the cellular structure, although the moisture level of the finished product will also influence the texture.

The range of extruded products is increasing almost weekly. Numerous possibilities of co-extruded products are being developed where fillings can be put inside an extruded shell. The variety of shapes and sizes is infinite in what is really the food product of the eighties.

Microwave cooking (dielectric heating)

An alternative heating mechanism to the conventional ones of conduction, convection and radiation, is available in the form of **dielectric heating**. Dielectric heating includes radio frequency and **microwave** cooking, best known in the microwave oven. The general principle of this type of heating and subsequent cooking depends on molecules in the food being electrically charged, in the same way as a magnet has a north and south pole. In an electric field a magnet, as shown by a compass needle, will attempt to orientate in a certain way to the field. In dielectric heating the electric field is reversed many millions of times a second. As a consequence the molecules, particularly water, have to re-orientate themselves to the change in fields many millions of times a second. Considerable heat is generated due to friction.

Foods containing water, therefore, heat up very quickly when subjected to dielectric heating. Foods placed in a microwave oven cook quickly because of this effect. Heat produced in the water brings it to boiling point, and this heat is then passed to other molecules alongside. It is generally believed that microwaves cook a food from the inside outwards. In fact, this is not true since all the food receives energy at the same rate, but the outside readily cools due to heat loss.

The most widely used example of dielectric heating is the microwave oven. First introduced in 1947 in the USA, the 'microwave' only started to become popular in 1972 in Britain. Modern lifestyles make this type of oven an essential kitchen appliance for many people. Rapid cooking of meals is possible, particularly defrosting or cooking frozen foods. Also, irregular meal times have increased the use of microwave ovens, mainly because of their speed and safety in use.

The demand for microwaveable products is increasing, and the Industry is developing ready meals and recipe dishes. However, existing products do not always cook well in a microwave oven. Fish, for example, may toughen to an unacceptable level when heated in this way, although fish such as mackerel are excellent when microwaved. Developing microwaveable products is difficult and it is not always economical to produce products which are solely for microwave ovens.

Some of the problems associated with microwave cooking are listed below:

(a) Microwave heating does not brown a food during cooking, so the attractive colour development or the flavour that goes with it do not occur. Late models have an additional heating element which browns the food in the normal manner

(b) The microwave heating process is difficult to control, with uneven heating and local hot-spots causing over-cooking. The speed of microwave cooking, although its main attribute, presents this difficulty in control

(c) Foods packed in aluminium cook much more slowly as the aluminium shields the food because of its poor penetration by microwaves. The aluminium sparks and arcs noisily. However, recent work has shown that a cardboard jacket around the aluminium food container controls this arcing and allows the product to heat up in a more controlled manner. Applications of this are being investigated for heating prepared meals on airlines

Microwave applications in the Food Industry are still fairly limited, the main ones being in *defrosting* and *tempering* frozen products, often prior to repacking. Other successful applications have been in the *drying* of pasta products and cooking of meat products. The system has also been used in certain *blanching* operations.

Considerable development and expansion is likely in this type of heat processing. It is an economical process, clean and space saving, with only moderate capital outlay.

Irradiation

For over 30 years **irradiation**, as a means of food preservation, has been investigated. It has been the most thoroughly investigated of all food processes. Yet, will the general public accept it without reservation in the same manner as frozen and canned products are accepted?

After the Chernobyl accident of 1986 and the threat of nuclear waste dumping in rural areas, the public associates *irradiation* with actual *radioactive contamination*. The latter is totally different from the radiations used in food irradiation, which can be either *electrons* produced in a linear accelerator or *gamma rays* from the radioactive decay of cobalt 60 or caesium 137. The electron beams can be directed but have poor penetration ability (less than 5 cm) whereas gamma rays, although non-directional, have considerable penetration ability.

The use of irradiation for treatment of food has been prohibited in the UK by legislation since 1967. In June 1989 the Government gave the go-ahead for irradiation, subject to future legislation. Since 1967 a number of significant reports have been produced in favour of irradiation which will undoubtedly lead to its legal acceptance in the UK. The Expert Committee of the FAO/IAEA/WHO (Food & Agriculture Organization (of the UN)/International Atomic Energy Agency/World Health Organisation) in 1981 proposed a general clearance of irradiation up to a dose level of 10 kGy (kiloGray). (10 kGy is equivalent to 1.0 Mrad which is sometimes used.)

A significant report was published in April 1986 by the Advisory Committee on Irradiated and Novel Foods which concluded that the irradiation of foods proved there to be no significant disadvantages or risks to health. The Food and Drink Federation welcomed the report by stating that 'food processors have long recognized the potential benefits of food irradiation as a process'. In 1986 Britain's Chief Medical Officer said that *irradiation* was a way of *preserving* food and was a *clean, simple* and *effective process*. Nevertheless, the process is not viewed with enthusiasm by the 'man in the street'. Why?

It is perhaps what irradiation does in a food material that gives rise to concern:

- Irradiation is intended to kill micro-organisms, particularly pathogenic organisms. At the dose level proposed of 10 kGy the irradiated food would be comparable with a **pasteurized** product, as **sterility** would not be achieved, but pathogens and many spoilage organisms would be destroyed

- Irradiation has the ability to break chemical bonds and reduce molecules into smaller pieces. The question of chemical toxicity has been raised as a result of this. However, at the levels of irradiation used there is no risk of any poisoning or any residual radioactivity, another area of possible concern

- Reduction of nutritive value has been cited as a consequence of irradiation, particularly a loss in some vitamins. However, losses have been found to be comparable to established methods of processing particularly heat processing

One method of assessing the quality of a food, particularly a raw material, is to estimate the number of micro-organisms present. However, as irradiation can destroy micro-organisms but usually leaves the food unaltered, this has been raised as a possible area for mal-practice. Unfortunately, some mal-practice has already occurred. In 1986 a company was prosecuted for exporting prawns, with a higher than normal microbial load to Holland for irradiation. The irradiated prawns were re-imported with a corresponding low level of micro-organisms and treated as if fresh. The danger from this treatment is that some micro-organisms can produce dangerous *toxins*, which are capable of food-poisoning, and these toxins may not be destroyed by irradiation. There is obviously a need for some kind of test to show that foods have been irradiated and for the packaging to carry some indication of the fact that the product has been irradiated.

It is surprising that the main criticisms of irradiation have come from the anti-additive lobby since, in the few commercial applications in operation in other countries, irradiation has replaced the use of chemicals:

- Inhibition of sprouting, particularly of potatoes, at low levels of irradiation (0.1 kGy) has been undertaken successfully for some time in Japan and also to a lesser extent in Italy and Hungary

- A main application now permitted in ten countries has been the treatment of spices and herbs to reduce bacterials loads – levels of irradiation vary up to 10 kGy

- Control of insects in grain and some fruit crops has been quite successful, particularly in the USA

- Holland has pioneered a lot of work on irradiated foods, particularly the inactivation of pathogens in a range of foods, some of which have been subsequently frozen. Irradiated chicken is available by this process in Holland

- Fruits have been successfully treated on a fairly large scale in South Africa, eg mould growth in fresh strawberries has been prevented, to produce an excellent longer-keeping product

- A range of fruits of tropical origin respond well to irradiation, as the process controls mould and insect spoilage

- Through irradiation the rate of ripening is retarded in some fruits but in others it may be accelerated. (Both this and the process of insect control present commercial opportunities)

- Foods containing fat may become rancid when irradiated as the process will initiative chemical changes leading to the process of rancidity. Fatty fish is such an example, but white fish has been irradiated successfully

If irradiation is to become acceptable it must meet two main requirements. Firstly, it must be a *viable commercial operation* offering an alternative method of processing to manufacturers. Secondly, it must be *readily acceptable* to the general public. The former is more easily achievable than the second with the current public attitude to irradiation and nuclear matters in general. However, with widespread use of irradiation for the production of pre-packed medical supplies and medicines, it must take its place eventually alongside the other methods of food processing.

Food packaging

Packaging is necessary (a) to protect a processed or preserved food from recontamination by micro-organisms and by the elements, (b) to contain it, and (c) to identify the product. In the last few years there have been rapid advances in packaging technology, mainly in the area of plastics with their endless versatility for food packaging.

Rigid plastic containers

Rigid plastic containers have shown rapid growth recently as food containers and development work is going on to replace the metal can with a plastic one. Sweets and chocolates have been contained in clear rigid containers for some time. Rapid development in this area has also occurred in carbonated drinks packaging, mainly using PET (polyethylene terephthalate) bottles. These convenient bottles are rigid enough for large volumes with little risk of breakage.

Plastic film

Plastic films have numerous applications and recent improvements have provided more effective barriers to gases, moisture and odours. There has been considerable growth in applications, such as the 'bag-in-box' idea, cling-wrapping and heat sterilizeable pouches for use instead of cans.

Although there has been considerable development in the field of multilayer materials, particularly laminates, the majority of plastic film applications are concerned with five types of films:

1 *Polyethylene* ('Polythene')

   (a) *Low density polyethylene* (LDPE) is widely used as it is strong, a good barrier to moisture and it heat-seals easily

(b) *Linear low density polyethylene* (LLDPE) is a recent derivative of LDPE, whereby the side chains of the polymer are extended giving a number of advantages. A stronger film with greater barrier properties is produced

(c) *Higher density polyethylene* (HDPE) is much stiffer with correspondingly greater barrier properties to moisture and gases. Because of this stiffness, very thin films can be produced, which are more resistant to higher temperatures than LDPE

2 *Polypropylene* (PP) is an even better barrier to moisture and gases. A number of versions of this are produced with better heat-sealable properties, greater slip levels for rapid mechanical packaging and opacity when required

3 *Polyvinyl chloride* (PVC) is a very good barrier to gas, but it is not such a good barrier to moisture as polypropylene. In very thin sheets it is used extensively for shrink wrapping of standard retail trays of meat, fruit and vegetables

4 Flexible versions of *PET* (see above) are making in-roads into a number of PVC applications

5 *Regenerated cellulose films* ('cellophane') have been used for some years as they have high burst strength, good barrier properties to fat and odour, and great clarity. When dry they are excellent barriers to gases, but this is lost when the film becomes wet. Many coated versions are now available which have overcome this problem

There are numerous combinations of films: as co-polymers and as laminates incorporating foil. These have been developed for particular purposes. Metallized films are a large growth area in this type of material and combinations of LDPE, foil and cellulose are common. These laminates have low permeabilities to moisture and gases but also exclude light which can accelerate deterioration in some products, as the uV fraction accelerates rancidity in fats.

The actual choice of a packaging film depends often on the properties of the food. Dried foods must obviously have packages with good moisture barrier characteristics. Foods containing fat must be packaged to stop the entry of oxygen, which might cause rancidity. Fresh meat, on the other hand, must gain oxygen to keep its bright red colour. Foods with high moisture content, for example fresh vegetables, must be packaged to retain as much moisture as possible but not to allow moisture droplets to accumulate on the inside of the package.

There are numerous packaging applications and these are being extended by the use of **modified atmosphere** (MA) packaging, where gas flushing is used, and by **aseptic packaging** techniques.

Glass

The highly effective and versatile flexible films cannot totally replace traditional materials such as *glass*. Glass is available in an infinite range of shapes, which can be produced more easily than their plastic rivals. Glass can be filled easily and presents an excellent barrier to moisture and gases. Lighter and stronger glass has been perfected with the advantage of being able to produce better shapes. Also, costs have been reduced so that many bottles are non-returnable.

The above has been only a brief review of food packaging, which is an enormous subject in its own right. The developments in packaging technology go along hand-in-hand with developments in food processing and even facilitate the widespread acceptance of new processed foods.

Range of products in different forms of packaging

4 Food spoilage and food poisoning

Foods, as well as meeting the nutritional requirement of humans, also meet the nutritional needs of a vast range of micro-organisms. Given the right conditions, micro-organisms will multiply rapidly in food and usually produce changes in flavour, aroma and texture. At the same time, a more sinister population of micro-organisms might become established in the food without any noticeable change in the organoleptic properties of the food. These organisms might cause an outbreak of food poisoning when the apparently wholesome food is consumed.

What are micro-organisms?

Micro-organisms are microscopic living things, mainly belonging to the plant kingdom, but some, such as protozoa, are animals. The main groups of micro-organisms are the *bacteria, moulds, yeasts, algae* and *viruses*. The one common aspect of the simple organisms is that they are small. Bacteria are often around $1\,\mu m$ ($1/1000\,mm$) across, yeasts are somewhat bigger at about $10\,\mu m$ and moulds can be larger as they form long thread-like structures.

Three of these microbial types are of significance in foods and these are *bacteria, moulds* and *yeasts*:

Bacteria – are extremely widespread in occurrence and cause food spoilage and some groups food poisoning
Moulds – are very common with their spores floating in the air ready to fall on and colonise a suitable food. Some moulds can produce toxins which cause disease
Yeasts – spoil foods, particularly those of a higher sugar or salt content

In this chapter it is the bacteria which are the most significant and further discussion will be concentrated on them.

The growth of bacteria

The multiplication of micro-organisms is generally referred to as 'growth' and as numbers of bacteria increase we say they are 'growing'. This does not mean the bacteria are increasing in size, only in numbers; since they can divide rapidly, their population can sometimes double in about 20 minutes.

(a) The growth of bacteria is rather slow at first as the organisms become established in the food; this is the *lag phase*

(b) Once established the bacterial cells start to divide ever more rapidly – the *log* or *exponential phase*. Sometimes at this stage the food might not show any outward signs of the bacterial population, although often obvious odours will be detected

(c) As nutrients are used up in the food, the rapid rate of multiplication slows down and the number of cells being produced matches the number of bacteria dying. This is the *stationary phase*, when there is a constant level of micro-organisms

(d) After a while the numbers dying exceed new cells and the growth cycle passes to the *death phase*. At this stage the food may well be 'off', but it may still be a good source of nutrients for a different group of bacteria which thrive in the conditions produced by the previous groups. In this way different groups of organisms can use the food as a source of nutrients until it is completely decayed.

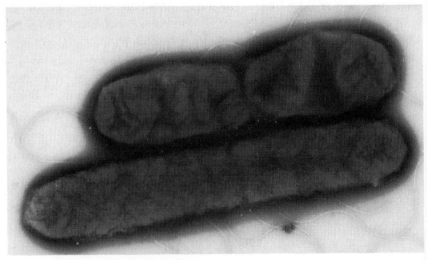

Enteropathogenic Escherichia coli, *isolated from a case of infantile diarrhoea*

Factors influencing microbial growth

Bacteria will not grow on merely any food, under any condition. There are a number of factors which influence the growth of bacteria (but there are often quite large degrees of tolerance):

- TEMPERATURE: although many bacteria can grow at widely differing temperatures, they often prefer to grow within a specific temperature range. Organisms which grow most readily at around room temperature and upwards (20–40°C) are known as *mesophiles*. This group contains many food poisoning and spoilage organisms. Some organisms actually grow more rapidly at refrigeration temperatures (0–10°C), this group contains the *psychrophiles*. Conversely, the *thermophiles* have an optimum at between 50°C and 70°C. The former cause spoilage in chilled foods, while the latter can withstand heat to quite high levels in cooked or processed food

- MOISTURE: Micro-organisms require differing levels of moisture in order to grow in foods (see Chapter 3). Most bacteria require high moisture levels, whereas most moulds can grow under low moisture conditions, for example on old leather. Some yeasts are capable of growing in high sugar products such as confectionery or salt products like pickles. A few years ago several million fondant-filled chocolate eggs exploded because they had become infected with a yeast which caused fermentation of the filling producing gas

- pH CONDITIONS: Organisms have a preference for certain pH conditions, many prefer near neutral but some, particularly moulds, can grow under very acid conditions.

- AIR: Some organisms can only grow in the presence of oxygen (or air) and are known as *obligate aerobes*. Those organisms which can only grow in the absence of oxygen are *obligate anaerobes*, whereas those which are capable of either are *facultative anaerobes*

In order to grow in a food the factors an organism requires for growth must be met. Some organisms are extremely demanding in their growth requirements, others are easily accommodated. Unfortunately, human food meets the requirements of most organisms without much difficulty.

Food spoilage and preservation

Micro-organisms eat our food in order to grow in much the same way as we do. They produce waste-products (sometimes toxic) and cause chemical changes in the food product. These changes can be detected as odour, flavour, colour and texture changes. Some are very noticeable and obnoxious, indicating that the food is 'bad'. Many of the changes produced by spoilage due to micro-organisms are obvious, like the rotting of fruit, souring of milk, putrefaction of meat and 'hairy' mould growth. Most of these foods are termed *perishable*, whereas foods like flour or spices are *stable*.

Whether or not spoilage occurs depends on a number of factors: (a) An initial load, perhaps of several million organisms, may be necessary to initiate spoilage; (b) the previously-described growth factors have to be met in the food; (c) the storage conditions under which the food is kept also affect the rate of spoilage.

In the processes of food preservation, the growth factors of organisms are controlled so as to prevent or retard their growth:

1 The temperature of storage may be reduced below the optimum growth temperature of most organisms present in the food. Thus in chilled foods only psychrophiles can grow

2 High temperatures may be used to kill all organisms present, for example in canning

3 In the process of drying, water levels are reduced to below the moisture level requirement of organisms so they cannot grow

4 Products can be made more acid, in pickling or fermentation, so that the pH of the product is outside the growth range of the organisms

5 Food can be packed in a vacuum or gas flushed to prevent oxygen supporting the growth of obligate aerobes

6 A range of substances, preservatives, can be added to inhibit the growth of micro-organisms

Food spoilage can occur in a product to produce noticeable spoilage so that people do not usually consume the product. However, some organisms can grow or be carried in a food without altering the food. These may then cause food-poisoning or some other disease.

Food poisoning

The number of cases of food poisoning caused by micro-organisms is increasing annually. In addition to the greater number of outbreaks, more groups of organisms are being recognized as capable of causing illness. The public is aware of food poisoning and will report cases more readily, rather than accepting their illness as another 'bilious' attack. There are greater numbers of small food outlets, cafés and fast-food establishments. Although many have very high standards, there are enough whose hygiene standards are below par to cause an increase in the outbreaks of food poisoning.

The presence of large numbers of micro-organisms in a food does not necessarily mean the food is a health risk. However, high numbers often imply poor hygiene standards at some time during the handling and preparation of the food. This poor hygiene could mean that the food has become contaminated with food poisoning organisms. For an outbreak of food poisoning to occur the food must become *contaminated with food poisoning organisms* and these must *grow to sufficient numbers* to cause illness. Some organisms produce toxins, sometimes very resistant to heat, which cause the poisoning and not the organisms themselves.

Contamination of foods by micro-organisms can occur from a variety of sources, not least humans. The air, soil, skin, animal feed and utensils can contaminate exposed foods. The contamination need not occur directly, but via a number of different vehicles. To illustrate the case of contamination of foods by micro-organisms a number of short case histories will be discussed.

Example 1

A common organisms found on the skin, in the nose, and in spots or boils is capable of growing in many foods and producing a toxin which causes food poisoning. This organism is called *Staphylococcus aureus*, or *Staph. aureus* for short.

A person preparing a trifle for a reception had a boil caused by *Staph. aureus* on the arm, but the plaster covering the boil came

59

away. Whilst preparing the trifle the person had touched the boil and then the trifle ingredients. The trifle was now contaminated with *Staph. aureus*, which grew in the trifle until it was consumed at the reception. After about three or four hours people who had been to the reception were feeling dizzy, suffering from nausea and vomiting. Some older people were liable to become unconscious. When the toxin had been expelled from their bodies most people recovered within a day. Obviously, this was a case of poor hygienic and the boil should have been covered and not touched. Correct washing of hands and hygienic preparation of the food was essential.

Example 2

A very serious type of toxin food poisoning is caused by the organism *Clostridium botulinum*. This bacterium produces very heat-resistant spores.

In the 1930s and 40s home-canning was popular in the USA and, to a lesser extent, in the UK. Home produced canned vegetables, meat and fruits were popular. However, on occasions products were insufficiently heat treated to sterilize the can contents. This was often carried out, after sealing the can, in a pressure cooker. Most organisms in the food would be destroyed by the heat but the spores of *Cl. botulinum* are more heat-resistant and therefore survived the process on occasions. The heat process stimulated the bacterial spores to germinate and to grow in the food in the can.

As the bacterium is an obligate anaerobe, conditions in the can of little or no oxygen are ideal. The bacteria multiply in the food and produce a toxin, neither of which can be detected in the apparently wholesome food. After eating the food, symptoms of food poisoning develop perhaps twelve hours later. The toxin is one of the deadliest poisons known and affects the central nervous system leading to paralysis and often death.

Example 3

The salmonella group of bacteria is well known for its food poisoning activities. In addition to causing food poisoning, it contains organisms capable of causing serious illness such as typhoid (*S. typhi*). The members of this genus are generally named after the place where they were first discovered. So names such as *S. london, S. newport, S. dublin* and *S. montevideo* are commonly encountered.

Salmonella causes food poisoning by infecting the gut of the victim and then multiplying inside the person. Irritation of the gut lining occurs and toxic by-products are produced leading to vomiting and diarrhoea. These symptoms vary from fairly mild to very severe resulting in death.

Salmonella can contaminate foods by various means:

(a) The bacteria may be carried by rodents, insects or birds which contaminate food directly, food preparation areas, or animal feed

(b) Infected animals can pass the organisms to humans if the infection is not detected or eliminated in cooking or processing. It is well known that chicken must be thoroughly cooked with no red meat to ensure the elimination of salmonella

(c) The most difficult form of contamination to trace is the human carrier. This is a person who does not show symptoms of a salmonella-type infection but has large numbers of the organism within the intestinal tract. Poor personal hygiene, particularly after using the toilets, can lead to contamination of food by the carrier's salmonella. Stool-testing of all personnel involved in the food operations is necessary to trace the carrier. The person should not be allowed in the factory or café until after at least two clear stool tests.

As the organisms need to grow in the gut before symptoms of food poisoning begin to occur, it may be several hours or days before an outbreak is discovered. In the case of some food-borne diseases, such as typhoid, very few organisms can cause the illness but then the incubation period may be two or three weeks. Obviously, to trace back to find the contaminated food is comparable to a murder investigation!

Example 4

Another type of food poisoning can occur in reheated meats. The organism *Clostridium perfringens* can contaminate meat during slaughter. Like *Cl. botulinium* the bacterium produces very heat-resistant spores which can survive the cooking of the meat. If the meat is stored for a time, the spores may germinate and the emergent bacterial cells will multiply. If the meat is only reheated and not thoroughly cooked again, the bacteria will survive and then increase in numbers. This occurs more rapidly in canteens with heated display counters. The contaminated meat, when consumed, releases its bacteria in the gut, where they produce a toxin resulting in food poisoning.

Other examples

Other groups of bacteria have been implicated in food poisoning from time to time:

1 Another spore-former, *Bacillus cereus*, has been known to cause food poisoning in boiled rice and in some shellfish

2 Bacteria shaped like commas can cause disease. A food-borne disease which has caused many deaths is of this type, and is known as *Vibrio cholerae* (or *V. comma*). As the name suggests, it causes cholera

3 A relative of the above bacterium has been implicated in food poisoning in some fish products, particularly in Japan. The bacterium is called *V. parahaemolyticus*

4 Recent work has shown that the genus *campylobacter* can cause gastroenteritis. The species particularly involved is *C. jejuni* which is found to contaminate foods of animal origin on occasions.

This is a brief look at the vast subject of microbial food poisoning (see also page 64). Unfortunately, the names or organisms are complicated but are in universal usage and so cannot be avoided. Food poisoning can also occur due to chemical contamination of foods, either deliberately or by accident. The contamination of cooking oil in Spain a few years ago was an example of this. However, food poisoning can and must be avoided and there is no excuse for poor hygiene in manufacturing or handling of foods.

Food hygiene

The main regulations relating to hygiene are the Food Hygiene (General) Regulations of 1970. These cover any food business, including shops, cafés, canteens, clubs, boarding houses, schools and factories. Anyone involved in handling food is subject to the

Regulations. They also apply to all foods and drinks and also to any ingredient:

Food premises

- The premises used in the food business must be clean and in good condition. The position and construction of the premises must not allow contamination of food in any way. The floors, walls and ceilings must be kept clean, and so must every toilet facility provided
- There must be an adequate, constant supply of clean, potable water, both hot and cold. Sanitary conveniences must be clean, with a 'Now Wash Your Hands' notice. Food handlers must have adequate hand-washing facilities, with soap, brushes and drying facilities (hot air drier or paper towels)
- A first aid kit must be readily available, with waterproof, brightly coloured plasters
- Staff should be provided with clothing lockers
- Equipment used for processing and handling must be in good condition and clean. Containers should be non-absorbent and capable of easy cleaning

Hygienic practices

- Provision must be made for separating and disposing of unsound food in a different part of the premises
- Food placed in the open for sale must be screened or covered if any risk of contamination is likely
- Foods intended for immediate consumption require special precautions, particularly meat, poultry, fish, gravy and imitation cream, and any product prepared from them. These foods must not be kept between temperatures of 10 and 62.7°C; consequently they must be cooled or kept really hot. Food poisoning organisms multiply rapidly at lukewarm temperatures
- Everyone involved in the food business should do all in his/her power to protect the food from contamination. Cleanliness is of utmost importance at all times:
  (a) Food must not be placed where there is a risk of contamination. This means that it must be kept at least 45 cm from the ground

(b) Food must not be packed in packaging materials that are not clean. Newspapers can be used as an outer wrapper for some products

(c) Animals, poultry or animal feed should not come into contact with food or packaging materials

Personal hygiene

- There are five rules for personal cleanliness to prevent micro-organisms from entering food:

    (a) Cleanliness of hands or other parts of the body in contact with food is very important. Hair nets should be used as hair can carry bacteria

    (b) Clothing and protective overalls must be kept clean and changed often

    (c) All cuts and grazes should be covered with a waterproof, brightly coloured dressing

    (d) Food handlers must not spit

    (e) Food handlers must not smoke while handling food or while in a room containing food

- Personal health is important and employees must report cases of diarrhoea, vomiting, septic sores and boils, discharges from the ear, nose or eye. 'Carriers' of diseases must declare the fact if known

- Serious diseases should be reported to the Medical Officer of Health and include typhoid, paratyphoid, salmonella infections, dysentery and staphylococcal infection

Most of these points are obvious and are routinely carried out or avoided, as the case may be, by competent food handlers. Failure to comply can lead to prosecution or perhaps, worse still, to an outbreak of food poisoning resulting in total loss of future business.

Current microbial problems

In 1988 Edwina Currie, then junior Health Minister, made her now infamous remark that 'most of the egg production in this country, sadly, is now infected with salmonella'. This resulted in an immediate drop of 50 per cent in egg sales, which subsequently levelled off to 10 per cent below normal.

Salmonella

In fact, the **salmonella** infection in eggs resulted from a type of contamination not encountered before. The organism involved is very specific and is called *Salmonella enteritidis* (phage 4). The bacterium infects the ovaries of chickens and, from there, gains access to the yolk of eggs. Only the white of the egg has defence mechanisms against bacteria, not the yolk. Thus, the consumption of under-cooked eggs, particularly 'runny' yolks and egg mousses, could lead to a salmonella infection. In 1988 there were about 10 500 cases of this type of salmonella infection out of 30 million eggs consumed daily.

Listeria

At about the same time as the egg crisis another organism, *Listeria monocyotogenes*, came to prominence. **Listeria** was found to contaminate 16 per cent of cook-chilled foods in a survey conducted by the Department of Health. For many years the organism had been known as a cause of abortion in sheep. Among humans, in 1988 there were 287 known cases of listerosis of which 50 of the sufferers died and 11 women had miscarriages. However, according to Sir Donald Acheson, the Government's Chief Medical Officer, 'one person in twenty has listeria in the gut and is feeling quite well'. Clearly, compared with more than 24 000 reported cases of food poisoning in 1988, the number of known cases of listerosis is very small. However, it should be realized that some minor cases of food poisoning may, in fact, be caused by listeria. It would therefore appear that the difference between a mild stomach upset and full listerosis depends on individual susceptibility: foetuses, babies, the very elderly, and those taking immuno-depressant drugs are most at risk.

Unfortunately, unlike most other bacteria, listeria can multiply rapidly if food is kept above a temperature of 4°C. Yet many supermarket display cabinets have been kept as high as 8 °C. This means that in such cases we cannot rely on the fact that spoilage organisms grow before food poisoning in chilled foods. The temperature is such that the food poisoning organisms rapidly multiply. As well as cook-chilled foods, salads, pâtés, and certain soft cheeses, including goat, have been implicated. The use of unpasteurized milk in cheese production will probably be banned throughout the EC as a result.

Botulism

Botulism is a rare but deadly form of food poisoning in the UK. There are several strains of *Clostridium botulinum* which produce toxins of varying potency in food, type A being the deadliest poison known. In June 1989 there was an outbreak of botulism in hazelnut yoghurt. The organism grew in a purée of hazelnut used for flavouring which was then added to the yoghurt in the final mixing. The pH of yoghurt is low enough to prevent *Cl. botulinum* from multiplying. However, the toxin is unaffected by the low pH and, therefore, can cause severe food poisoning. A total of 27 cases of poisoning occurred.

It is very easy to view these microbial problems superficially and generate, for example, 'listeria hysteria'. Although more cases of food poisoning are being reported, the actual number of cases may not be rising significantly. Reports of salmonellosis have shown an almost continuous annual increase since data were first collected in the 1940s. There is now a much greater public awareness of food poisoning and a willingness to report cases. Analytical and assay techniques have improved considerably so foods can be shown readily to contain food poisoning organisms. However, the number of organisms present may not be large enough to cause poisoning, since in most cases considerable numbers are required.

The current preoccupation with food hygiene is a mixed blessing. In earlier times people lived with a greater range of infection risks and children acquired immunity at an early age from low-level infection. Most people live in daily microbial risk with no ill effect due to their immunological systems. The removal of all bacteria, if that were ever possible, would compromise our immunity systems so that exposure to any organism would have a devastating effect.

section 2

Foods

In this section individual foods will be reviewed. Generally the foods will be considered from four aspects:

(a) composition
(b) processing
(c) nutrition
(d) effects of cooking

This section is intended for quick reference. Further details may be found in books dealing specifically with the individual foods. Terms or words of scientific origin are explained in the Glossary.

Dairy products

Milk

Composition

Milk is one of the most complex foods. Milk fat globules are dispersed with proteins in a solution of salts, sugar and vitamins. The salts include salts of sodium, potassium, calcium and magnesium with chlorides, phosphates, citrates and bicarbonates. All vitamins are present, but some in very small quantities. Vitamin C is reduced by heat processing and significantly by exposure to light. Direct sunlight can trigger **rancidity** in the fat and can cause a reaction between the amino acid methionine and the vitamin riboflavin, producing a cabbage-like flavour. These are good reasons in themselves for not leaving milk bottles on doorsteps.

The main sugar in milk is **lactose** which is only about one tenth as soluble as sugar (**sucrose**). This lack of solubility causes lactose crystals to be produced in some milk products such as ice-cream and sweetened condensed milk.

The fat globules contain butter-fat, fat-soluble vitamins, **lecithins** and **carotenoids**. The globules are surrounded by a complex membrane made of proteins, **phospholipids** such as lecithin, and enzymes. This membrane prevents the globules coalescing and rising to the surface. Any heat processing will damage the membrane and so globules may coalesce. Globules rise to the surface to form the cream layer, traditionally a measure of milk quality! Fat content of milk varies from about 3.4 per cent in Holstein to 5.8 per cent in Jersey cows. Butter-fat is unusual in having a number of short-chain fatty acids, and saturated fatty acids are common.

There are dozens of different proteins in milk which can be divided into those forming **curds** and those forming **whey**. The former react to acid and form curds by the action of **rennet** (see Cheese). The whey proteins are unaffected by rennet. **Caseins** make up the curd protein, and consist of mixtures of phosphoproteins (proteins containing phosphate). Some caseins (α_s) are sensitive to calcium and are readily coagulated by calcium in many forms. Another type of casein (κ) (kappa) is resistant to calcium and in fresh milk protects the α_s from precipitation by calcium. The caseins are bound together in microscopic bundles known as **micelles**, with the κ-casein on the outside protecting against calcium.

The whey proteins include albumins, globulins, enzymes and breakdown products. The protein β-lactoglobulin contains sulphur atoms which are exposed by heating milk above 74 °C, causing the protein to uncoil. This gives a cooked milk flavour, found in sterilized milk. The gas hydrogen sulphide is formed in this reaction with its characteristic strong smell. The main whey protein is α-lactalbumin which is associated with enzymes involved in lactose synthesis.

The composition of cow's milk varies according to the stage of lactation, season, weather, feed, time and period between milkings and breed. The latter is probably the most significant factor.

Processing

As milk is an excellent growth medium for micro-organisms it must be kept chilled, or processed to reduce or eliminate the organisms involved. Milk is capable of carrying not only spoilage organisms but also a range of pathogens. It used to be the main carrier of tuberculosis and until recently *Brucella* was quite common. This strain of organism causes abortions in cattle and an undulating 'flu-type illness in humans. Many other diseases can be carried in milk and groups of organisms such as salmonella and staphylococcus are common.

Pasteurization was developed during the last century by Louis Pasteur whilst working to prolong the life of some wines. Milk is heated below boiling point but at a temperature high enough to destroy harmful bacteria, in particular *Mycobacterium tuberculosis*, the causative organism of TB. Modern pasteurization processes use special plate-heat exchangers to heat the product quickly (71.7 °C for 15 seconds). The process is called **high-temperature-short-time (HTST) processing**.

Heat damage to the fat-soluble membrane necessitates the process of **homogenization** to prevent separation of the cream layer. At a temperature of about 60 °C the milk is forced under pressure through a small orifice. The fat droplets are reduced in size to about 1–2 μm (microns) (ie 0.001–0.002 mm). Small droplets do not coalesce very easily and so remain in a stable dispersion in the milk. This process is always carried out for carton milk.

Milk bottling line

Pasteurized milk, as it still contains micro-organisms, will become sour in a few days, although modern hygienic dairy processes have lengthened the shelf-life by at least two days. **Sterilization** is carried out to make the milk 'commercially sterile', which means that the milk is free of all micro-organisms, with the exception of a few spores. Traditional sterilized milk is processed in-bottle by heating the filled bottles to 120 °C for 15 minutes, then cooling to 75 °C by water sprays. The strong cooked flavour is either liked or totally disliked.

As the flavour takes some time to develop in sterilized milk, the modern process **UHT (ultra-high temperature)** can achieve sterilization without the flavour changes being so apparent. The process is a HTST process in plate heat-exchangers using a temperature of 132 °C for 1 minute. The 'long-life milk' is packed in cardboard cartons which have been sterilized by various methods, often using hydrogen peroxide. A more recent process achieves the same result by injecting steam under pressure into the milk. This process, known as **uperization**, is common in Europe.

Nutrition

Milk is one of the most nutritious foods. Unfortunately it has been criticized recently for its saturated fatty acid content. However, it is this fat that carries supplies of vitamins A and D. There has been a rapid growth in skimmed and semi-skimmed milk because of concern over butter-fat. The protein of milk is of high quality. Milk is a good source of calcium and is, therefore, a significant food in fighting osteoporosis later in life.

- *Pasteurization* alters the nutritional value of milk surprisingly little. The vitamin C content, which is only about 1 mg/100 g, is reduced by half. Vitamins B_1 and B_{12} are reduced by about 10 per cent
- *Homogenization* makes milk less stable to heat and more sensitive to the effects of light
- *In-bottle sterilization* of milk causes significantly greater losses in vitamins and a reduction in the nutritive value of proteins
- The *UHT* process has the same effect as pasteurization. However, during long storage of UHT milk dissolved oxygen causes severe losses of most vitamins

The effects of cooking

- When milk boils a skin is formed on the surface. This skin is a complex formed from casein and calcium, resulting from evaporation of water at the surface, causing the protein to become concentrated. This evaporation during boiling can be reduced by covering the pan, which also reduces the amount of skin formed
- Milk readily 'burns' on the bottom of pans as protein complexes, such as the casein micelles, sink and stick to the hot metal
- Cooking in the presence of acid causes curdling of the milk and a curdled appearance will be obtained if large molecules of starch or cellulose are added. These molecules act as sites on to which the protein molecules stick. The lactose and proteins react to form brown pigments, particularly in any skin that forms, and cause a change in flavour. At a higher pH, the reaction, known as the **Maillard reaction** is accelerated

Cream

Composition

The fat content of milk is around 3.5–4 per cent, so the amount of cream removed from milk will depend on its fat content. Cream separates gradually when milk is allowed to stand after pasteurization. However, commercially, cream is separated from the skimmed milk using a plate cream separator, a type of centrifuge.

The various types of cream available are given below, starting with the richest:

	Fat content (minimum %)
Clotted cream	55
Double cream	48
Whipping cream	35
Sterilized cream	23
Single cream	18
Sterilized half cream	12

Processing

The cream separator is a centrifuge built up of a number of plates. The separator rotates at speeds of about 6500 rpm and separates the heavier skimmed milk from the cream.

- Cream which is to be sold as fresh cream must be pasteurized at 82–88 °C for 10 seconds
- Clotted cream is made traditionally in the South West by heating milk to about 82 °C and allowing it to cool overnight in shallow containers. The cream is then skimmed off the milk
- New cream products have emerged in which part of the butter-fat is replaced with vegetable fat. It is necessary to use **emulsifiers** and **stabilizers**, such as locust bean gum and guar gum, to obtain a stable product. **Carotenoids** are added to give colour
- UHT cream is given the same heat treatment as UHT milk.

Nutrition

Formerly a luxury product, eaten occasionally, cream is readily available and affordable nowadays. However, it is very rich in fat, a significant amount of which is saturated fat – hence the development of cream substitutes which are supposedly less of a health risk.

The vitamins A and D are more plentiful in thicker creams than those of a lower fat content, and, because of its high fat content, the product has a high energy value.

Effect of cooking

Some chefs consider cream to be more useful than milk since, in the former, the protein content is diluted by the enormous amount of fat present. This means that, when cream is heated, a skin is not so easily formed (as when making a sauce) and it seldom leads to curdling in the presence of acid.

A main cooking application is the *whipping* of cream:

- Whipping can only be achieved with fat contents between 30 and 38 per cent

- During whipping air is beaten into the cream and the bubbles become stabilized. Protein molecules become denatured somewhat and orientate themselves around the air bubbles to form a film, thus protecting them. This operation is comparable to the action of an emulsifying agent surrounding a fat droplet in water

- The high fat content of the cream gives viscosity, thus preventing the bubbles rising to the surface and liquid draining through the foam produced in whipping. The fat globules tend to coalesce and stick to the protected bubbles, thus aiding their stability

- The fat globule membranes of homogenized cream are damaged and so do not coalesce and stick to the protected bubbles well. Consequently homogenized cream is of poor whipping quality

- Temperature is important: 7 °C or lower is best; above 21 °C very little foam can be produced

- Large fat globules produce stiffer foams than small fat globules – Channel Island breeds give the largest globules

- Over-whipping can cause fat globules to clump together, and the process of butter-making begins

Butter

Composition

Butter has been made for centuries by churning cream. Butter contains over 80 per cent fat, 16 per cent water (the legal maximum) and small amounts of protein and, in some varieties, salt. The fat is a mixture of liquid and crystalline fat, water droplets and some air bubbles.

The yellow colour of butter arises from carotene, which often comes from the cow eating grass. Spring or summer grass tends to give more carotenes in the milk fat and slightly more unsaturated fatty acids. This results in a softer butter from summer milk than from milk produced in winter. Blending of different types ensures a consistent product.

Processing

Both cream and butter are *emulsions*: cream being an emulsion of fat in water, whereas butter is an emulsion of water in fat. This 'inversion of a colloid' is achieved by the process of churning.

Although fresh cream is often used for butter-making, sometimes the cream is ripened using a **starter culture**:

- Special organisms, such as *Streptococcus cremoris* and *S. lactis* are added to the cream and allowed to grow to produce **lactic acid**, usually to a concentration of 0.25 per cent

- The cream is then pasteurized and cooled to around 10 °C, at which temperature churning is most successful

- Churning involves violently agitating the cream for up to about 30 minutes. As when being whipped, during this process air is incorporated into the cream and the fat globules associate around the air bubbles. However, the temperature is higher causing the fat globules to soften and coalesce

- Liquid fat cements the solid fat into small lumps of butter and any foam produced is destroyed by the free liquid fat. In this way the butter granules gradually grow and work together to form larger ones

- The butter milk is removed and the remaining butter is washed, sometimes salted and worked to obtain the desired consistency

Continuous buttermaking machine (Alfreton Creamery)

- The final water content must be below 16 per cent

Modern butter-making, like many food processes, is continuous. Cream is fed continuously into a horizontal cylindrical churn where churning occurs immediately. After draining off the butter milk, the butter is worked through sets of plates. On emerging from the churn, the butter is packed.

Nutrition

Butter sales have declined enormously because of the association of saturated fats with heart disease. Butter is a high energy product, 100 g giving 3140 kJ (750 kcal). However, it is also a source of the fat-soluble vitamins A, D and some E. Summer butter tends to be richer in vitamins than winter butter.

Effects of cooking

Butter gives a characteristic flavour to foods and is much used by the French. The wide melting range of butter-fat makes it very useful for the manufacture of a number of products, as it is essentially a high melting point fat, with melting occurring at around 36–40 °C. Such products include pastries, biscuits, fillings, sauces and cakes.

Butter melts when heated in a saucepan and begins to sizzle as the temperature nears 100 °C as the water bubbles off the melted fat. The proteins present turn brown as the temperature rises and may burn, imparting a harsh flavour. To guard against this, **clarified** butter, from which the proteins have been removed, can be used. The addition of cooking oil also helps.

Butter can become **rancid**, either through oxidation or hydrolysis, the latter being caused by micro-organisms, especially moulds. Salting was used originally to prevent the latter, but with refrigeration it is now not required. In fact, salted butter shows an accelerated rate of oxidative rancidity, and so does not keep as well as unsalted butter.

Milk powders

Composition

There are a range of powders produced from milk, skimmed milk and whey. Some of these can be used directly as milk substitutes or as ingredients in other food products.

(a) **Spray dried** milk powder is the major source of milk protein used as a food ingredient. All but a small fraction of the water is removed from the milk to produce a whole milk powder, usually using a spray drying technique. However, the mixture of the milk with air, and the contact of the butter fat with salts in the dried product quickly lead to a rancid powder. Skimmed milk powder is produced in a similar way, but the absence of fat gives a longer keeping product

(b) **Filled milks** are produced by adding vegetable fat to skimmed milk prior to drying. This gives a similar mouthfeel and flavour as ordinary milk on reconstitution with water. Such milk powders are popular within the Catering Industry

(c) **Whey** powders are used as alternatives for skimmed milk powder as they are cheaper. However, flavour and problems with gritty lactose crystals sometimes limit their use

Processing

For many years milk powders were produced by **roller drying**. During this process the milk is dried by contact with an internally heated roller. As the roller rotates water is driven from the product, leaving a flaky powder. The process is relatively long so heat damage to the product can occur (for example, browning due to the Maillard reaction is common). The nutritive value and reconstitution properties of the powder are also damaged.

Spray dryer from Alfreton Creamery

Spray drying is now by far the most common method of producing powders which are more soluble with good colour and flavour:

- The milk is generally concentrated in an evaporator prior to drying
- The concentrated milk is then passed through an atomiser which converts it into tiny droplets
- These droplets, on falling through the funnel shaped drier, meet a blast of hot air (180 °C) and dry instantly to produce a fine powder

Unfortunately, some powders are so fine that they float on top of water at the time of reconstitution. These powders have poor wettability and dispersability. Some products are **instantized** by rewetting the powder or not allowing it to dry completely. The powder particles tend then to stick together and form granules which are subsequently dried by a **fluidized-bed** system. The granules act like sponges during reconstitution readily absorbing water then dissolving in it.

Nutrition

Milk powders are excellent sources of high quality protein and there is generally little loss in nutritive value during drying. However, overheating, as in any process, will cause protein damage and vitamin loss. Browning due to the Maillard reaction also causes a reduction in nutritive value. The lactose and an essential amino acid, lysine, are thought to be mainly involved in this reaction. The Maillard reaction will occur slowly during storage of milk powders, particularly if the moisture content is allowed to rise to 5 per cent or above.

Effects of cooking

There are many examples of the use of milk powders in cooking and processing. The main uses are in batters and doughs for baked products.

- Milk powder can be added to encourage browning. For example, a small amount of milk powder accelerates the browning when producing caramel in sweet production
- The proteins in skimmed milk powder have useful properties in that they act as emulsifiers and water binding agents

- Whey powders have useful functional properties, particularly as they can be whipped to produce foams. Around 10 per cent of egg white can be replaced by whey proteins when producing meringues
- Whey powders are also of use in meat products, where they contribute to texture

Concentrated milks

Composition

Two types of concentrated milk are produced by evaporating off some of the water of whole milk.

(a) *Evaporated milk* has its water content reduced to about 70 per cent. It is then homogenized and preserved by sterilizing in cans. A poor granular texture is offset by adding sodium citrate, calcium chloride or disodium phosphate. The greater concentration of lactose causes some browning to occur which gives colour and flavour to the product.

(b) *Sweetened condensed milk* is produced from pasteurized milk that is subject to evaporation. A sugar syrup (60–65 per cent sugar) is added during evaporation so that the finished product may finish with a sugar content of around 50 per cent. This high sugar content preserves the product.

Processing

The evaporation process must be carried out under vacuum to reduce the cooked milk flavour, although this is still present to quite an extent. Temperatures can be as low as 55 °C during evaporation.

The addition of sugar to sweetened condensed milk tends to make the less soluble lactose come out of solution. If this occurs slowly in the cans of condensed milk, large marble-sized crystals of lactose are produced. Generally, during the later stages of evaporation, the condensed milk is 'seeded' with tiny crystals of lactose. This procedure makes the lactose within the condensed milk crystallize out quickly thus producing only very small lactose crystals which cannot be detected in the product.

Nutrition

As with all heat processes, there is some loss of vitamins, usually vitamin C and some of the B vitamins. The products nevertheless are highly nutritious and good protein sources. Recent developments for use in developing countries where there is little or no dairy industry have been the recombination of skimmed milk powder, milk fat and water to make these concentrated milks.

Use in cooking

Concentrated milks are useful as they supply milk fat and milk solids in a concentrated form. They can be used in many cooking applications to advantage instead of milk.

Evaporated milk can be whipped because of its higher viscosity and protein content. Whipping is improved by chilling and slight acidification with lemon juice. Traditionally evaporated milk has substituted for cream in many applications.

Ice-cream

Composition

Ice-cream has been produced in some form or other since the 16th century, well in advance of the advent of refrigeration, although the type commonly found today probably started in Paris in 1774. Ice-cream is an emulsion of fat in liquid, in which has been dissolved or suspended a number of particles and which is then frozen. In the frozen state ice-cream contains ice crystals, air bubbles, fat globules, proteins such as casein, stabilizers, sugar, flavours and colour.

A soft-serve ice-cream contains perhaps 6 per cent fat, 11 per cent skimmed milk powder, 14 per cent sugar, 1 per cent stabilizer and 68 per cent water. A hard ice-cream contains more fat, probably 12 per cent. Legally ice-cream must contain at least 5 per cent fat and 7.5 per cent solids-not-fat. Ice-cream mixes must be a careful balance of ingredients. Too much milk powder can lead to sandiness as lactose may crystallize from the powder. Too little milk powder will mean that water is free to form large coarse ice crystals. Stabilizers and emulsifiers such as **gelatine**, **alginates**, **pectins** and carageenans are used to improve the body of the product, to give a smooth texture by inhibiting large ice crystals, and to control the thawing characteristics.

Processing

There are three stages in ice-cream manufacture: preparing the mix, freezing it and hardening it.

(a) *Preparing:* The ingredients are mixed together and heated to dissolve the sugar and to pasteurize the mix

(b) *Freezing:* The freezing process may be carried out after homogenization. During this operation the mix is stirred quite vigorously to incorporate air.

   The volume of the mix after freezing should have increased by 100 per cent, this is called an 'overrun of 100 per cent'. Lower overruns give a tougher, less palatable product. Rapid freezing is necessary to ensure only small ice crystals are produced, but, because of the dissolved sugar and other ingredients, some of the mix always remains unfrozen.

(c) *Hardening:* During the hardening stage more of the unfrozen water is removed and deposited on to the ice crystals which grow in size. The result is that the ice-cream becomes harder as the ingredients bind closer together.

Compared with hard ice-cream, soft-serve types have less air mixed into them with overruns of about 50 per cent. Hardening is not carried out as the product is usually served directly from the freezer.

Nutrition

The 'health record' of ice-cream is excellent:

- It is a good supplier of energy because of its sugar content
- The fat content of ice-cream is predominantly saturated, even in cases where vegetable fats are used
- Serve-from-the-fridge types contain more unsaturated fats

Use in cooking

The storage of ice-cream can present problems since ideally it should be kept at $-18\,°C$ or below. As with other frozen products, fluctuating temperatures result in small ice crystals (which are unstable) readily combining to make large ice crystals even larger.

Thus refreezing ice-cream in a domestic freezer produces very large crunchy crystals.

Hard ice-cream can be used in a range of products, such as arctic roll. Ice-cream sundaes are usually made from soft ice-cream, although a harder variety may be used on occasions.

Cheese

Composition

Cheese has been produced since Biblical times and is a surprisingly consistent biological product. It is made from milk to which is added a starter culture of lactic acid producing bacteria. As the pH falls **rennet** is added to curdle the milk. The curds are then eventually converted into cheese.

In effect, cheese is a mixture of proteins (the **caseins**), milk fat, water and salts. The moisture level varies from variety to variety and there are many hundreds of different types. In fact, it is during **ripening** that the cheese develops its individual character.

A number of reactions occur during ripening:

* The proteins are hydrolysed in part by enzymes to liberate amino acids which contribute to flavour. Very soft and creamy cheeses are produced by almost complete hydrolysis of the proteins
* Fats are similarly attacked by enzymes to liberate fatty acids. The short-chain ones, such as butyric acid have a strong flavour
* Sometimes moulds are added, as in blue cheese, and these attack the constitutents further to produce strong flavours and odours like ammonia

Processing

Cheese making varies for each type produced but the general principles found in Cheddar cheese manufacture apply to most others. *Pasteurized* milk is used, which, in some countries, does not just come from cows but also from goats and sheep. Pasteurization is necessary to remove any contamination and to prevent food poisoning, especially in the higher moisture content soft cheeses. Pasteurization does, however, affect the finished cheese. It has been found to retard the action of rennet and to slow down ripening by inactivating enzymes. The differences compared with cheese made from fresh milk are nevertheless acceptable as there is a preference for milder cheese.

Cheese vat with starter and rennet

(a) The first step in cheese making is the addition of the **starter culture**, usually the bacteria *streptococcus lactis* and *S. cremoris*. These cultures are very susceptible to **bacteriophages** which can render them impotent

(b) The milk is held at about 25 °C as the culture of bacteria produces *lactic acid* to a concentration of about 0.17–0.2 per cent

(c) At this level of acidity **rennet** can be added. Rennet is a preparation of the enzyme *rennin* or *chymosin*, which is traditionally extracted from the fourth stomach of calves. It is now available from microbial sources

(d) The rennet attacks the casein micelle by splitting the κ-casein very precisely in one particular place in the protein's chain of amino acids. The protective effect of the κ-casein is destroyed so the calcium-sensitive $α_s$ casein is exposed. This fraction combines with calcium in the presence of phosphates and some citrate to form the *curd*

(e) The *curd*, now a solid congealed mass, is cut into cubes and the *whey* is allowed to drain away

(f) The temperature is increased to about 40 °C to make the curd contract and to expel further whey

(g) Again the cubes of curd coalesce into a solid mass which is cut into slabs, about 30 cm by 60 cm

(h) These slabs are piled onto each other and turned every 15 minutes for about two hours. This is known as 'cheddaring' and helps to develop the desired texture of the cheese

(i) After cheddaring, the slabs are milled into small pieces and about 0.1 per cent of salt is added

(j) The pieces are then put into a mould and pressed for one or two days

(k) This 'green cheese' is covered in cheese-cloth or plastic and ripened for several months at about 10 °C

Cheese varieties

The Cheddar process can be modified to produce a range of cheeses of differing acidities and hardness.

- Harder cheese is produced by
  (a) heating ('scalding') the curd to a higher temperature;
  (b) cutting the curd more finely; or
  (c) exerting greater pressure during processing
- Blue cheeses are not pressed but loosely packed into moulds and inoculated with needles dipped into cultures of moulds, particularly penicillium such as *P. roquefortii*
- Very soft cheeses are produced by the action on the cheese proteins of enzymes from moulds such as *P. camembertii* used to make Camembert and Brie.

Processed cheese

Processed cheese is a manufactured product from a number of cheeses that are emulsified with green cheese, emulsifier and water. The emulsifiers are salts, usually phosphates and citrates of sodium, potassium or calcium. The cheese is milled into small pieces and heated. It is then wrapped in foil containers.

Spreads are produced from processed cheese by adding more water and stabilizers such as gums or gelatine to give a smooth texture.

Nutrition

Cheese contains most of the components of milk and is a highly nutritious food:

- It is a good supply of protein (mainly casein) and calcium, containing about three-quarters of the protein and calcium content of milk
- All the fat content of milk (which is largely saturated) is found in cheese
- The fat soluble vitamins of milk are therefore also present in cheese

Effects of cooking

Again it is the proteins in cheese which are significant during heating. The 'stringy' nature of cheese during cooking is due to the *coagulation* of *casein* which separates from the fat, water and other constituents. During ripening the casein is broken down by enzymes, so well-ripened cheese is not as 'stringy' during cooking as mild or soft cheeses. As a result of this, hard, ripened cheeses disperse better into liquid when making a cheese sauce.

Overcooking of cheese dishes should be avoided as the cheese proteins become hardened and indigestible.

Yoghurt

Composition

Like cheese, yoghurt is a fermented milk product. It can be made from whole or skimmed milk, and often dried milk powder is added to provide additional casein to form a firmer gel. The starter culture used betrays the origin of yoghurt: *Lactobacillus bulgaricus* and *Streptococcus thermophilus*. Yoghurt was probably discovered by chance in Biblical times. It is very popular in the Balkans but is now widely consumed throughout Europe and America.

Processing

There are two types of yoghurt, set and stirred: (a) *Set yoghurt* is produced by allowing the fermentation and gelling to occur within the container in which the yoghurt is sold; (b) *Stirred yoghurt,* the most popular type in the UK, is fermented in bulk then packed

- The majority of yoghurts are low fat, so the fat is separated from the milk in the same way as in cream production
- The solids content of the separated milk then has to be increased by adding milk powder or evaporated milk, and sometimes some of the water is evaporated from the skimmed milk
- To the milk base are added ingredients such as sugar, stabilizers, colours and flavours
- The mix must then be homogenized and pasteurized
- After cooling the starter culture is added
- Incubation takes place at 43 °C for at least 1.5 hours (sometimes up to 4 hours)
- The warm yoghurt is poured into containers and kept warm until fully coagulated
- The product is then kept chilled until consumed, usually within 14 days

Stirred yoghurts can be made by continuous manufacturing processes, and are fermented with continuous stirring to prevent curd formation. Fruit and syrup is put into the familiar containers followed by the yoghurt mix. The product is then chilled and distributed.

There have been a number of developments in yoghurt manufacture, including frozen yoghurt and long-life varieties, which are comparable to UHT milk for keeping qualities.

Nutrition

The food value of yoghurt is similar to milk.

- It is a good source of protein and calcium, particularly if extra milk powder has been added
- Low fat yoghurts will be correspondingly lower in fat-soluble vitamins, so some manufacturers add these as supplements
- There is a tendency for more people to prefer natural yoghurt without colours or additives – these are generally set yoghurts
- Surprisingly, the quickest growth of yoghurts is in Greek style

products with fat levels of 6–10 per cent, the former usually being made from sheep's milk

- Yoghurt has, for centuries, been associated with health giving properties. It has been claimed that the starter organisms populate the gut and prevent the growth of harmful micro-organisms. This might not be the case since the organisms do not survive in the intestines. However, the bacterium *Lactobacillus acidophilus* is sometimes used in the starter culture and this, it has been claimed, can help patients recover from gastroenteritis.

Use in cooking

(a) Yoghurt can be used in a similar way to cream but has the advantage of being acidic and having some flavour

(b) It can be used to advantage in sauces, salad dressings and soups

(c) Yoghurts are also gaining popularity for use with breakfast cereals and in lunch boxes

Eggs

Composition

The majority of eggs consumed are hens' eggs, and occasionally those from ducks, quail, gulls and guinea foul. An egg can be divided into three main components: the outer shell, the white and the yolk.

(a) The *outer shell* of an egg is composed of calcium carbonate and has many small pores used for gas exchange. The pores are covered with a cuticle made of wax which protects against water-loss and prevents micro-organisms entering the egg

(b) The *egg white* is a mixture of proteins, some of which have remarkable properties. The white is thicker around the yolk to anchor it in place, with the help of strands of a protein called *chalazae*. A number of the proteins protect against microbial attack such as the protein *lysozyme* which causes bacterial cells to split open. Some proteins remove nutrients required by bacteria for growth, for example *conalbumin* binds up iron and **avidin** combines with the vitamin **biotin**. The protein *ovomucoid* inhibits the action of the enzyme trypsin which breaks down protein. These properties are lost in cooking the egg

(c) Egg yolks are rich in nutrients with several different proteins, including phosphoproteins and lipoproteins. There are also quantities of lecithin which, together with some lipoproteins, make egg yolk a very good emuslifier

Eggs are one of a very few foods where the pH can actually be above pH 7, sometimes even as high as 8 or 9. This high pH again has an anti-microbial function.

Processing

The majority of hens' eggs are produced in the battery system, often with automatic feeding and egg collection. There is a growing preference of free-range eggs, particularly the brown types.

Eggs are sent to a packing station, where they are **graded** and inspected. There are seven weight classes, Size 1 being 70 g or over, reducing by 5 g to size 7 which is 45 g or under.

Frozen whole egg is produced for the processing industry. Eggs are cleaned and broken mechanically. They are then homoge-

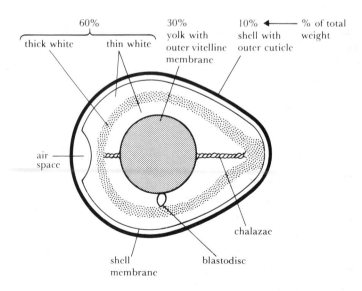

Structure of an egg

nized and pasteurized (63 °C for 60 seconds). Fragments of shell are filtered out and the liquid egg is filled into cans and frozen. *Dried egg* is produced in a similar way, but the egg is usually dried by spray drying.

Freeze-dried egg is a high quality product but its high cost generally does not justify its use.

Nutrition

- It is a mistaken belief that brown eggs are better nutritionally than white. Both are excellent sources of protein, vitamins A, D, B complex and B_{12}, and riboflavin

- The yolk is about 34 per cent fat, 16 per cent protein and 50 per cent water. Unfortunately about 5 per cent of the fat is cholesterol. This has led to advice about restricting the number of eggs eaten to perhaps three per week. However, the amount of cholesterol absorbed into the body from eggs may not be as great as originally thought, and so eggs have been reprieved somewhat on this account

- The yellow colour of the yolk is due to **xanthophylls**, carotenoid derivatives. However, unlike a number of carotenoids these cannot be converted into vitamin A in the body

Use in cooking

Eggs are unique in the range of properties they exhibit which are of value in cooking:

(a) The yolk has excellent emulsifying properties and stabilizes emulsions, such as mayonnaise, easily

(b) The yolk can also thicken liquids to form solids

(c) The egg white has excellent foaming properties which are the basis of cakes and meringues (see below)

When air is whipped into egg white, a foam is produced which is stabilized by the globulin proteins, which comprise about 10 per cent of the egg white proteins. These proteins have the ability to uncoil and form a film around the air bubbles as they are whipped into the egg white. This film is like an emulsifier around a fat droplet in water, it stabilizes the air droplet and prevents it separating from the mix. When the product is cooked the protein is denatured completely and retains a foam structure, as, for example, in meringues. Any trace of fat, even the egg yolk, will interfere with the development of the film around the air bubble. It will then be impossible for the egg protein to stabilize the foam produced by whipping. Plastic dishes tend to retain some fat on their surface and so should not be used when whipping egg whites.

Boiled eggs sometimes show a grey discolouration around the yolk, usually when the eggs are older with a high pH. This is due to the production of iron II sulphide, with the iron coming from the yolk and sulphur from some proteins such as ovalbumin. On heating, the sulphur is liberated from the protein as hydrogen sulphide gas (the smell of cooked or bad eggs) which reacts with iron. Cooling rapidly in cold water minimizes the effect.

Non-dairy fats

Margarine

Composition

The first margarine was made in 1869 by a French pharmacist, Hippolyte Mège-Mouriés, using beef tallow, skimmed milk and minced cows' udder! In many countries margarine caught on quickly as a cheap butter substitute. However, resistance from the dairy industry resulted in the USA in margarine being defined as a 'harmful drug' with restricted licensed sales.

Modern margarines are blended, and refined vegetable oils are so carefully controlled that the final product is always the same. The process of **hydrogenation** has enabled vegetable oils to be converted into hard saturated fats. Different combinations of saturated and polyunsaturated fatty acids can be used to produce margarines of different degrees of softness and plasticity. Soft-spread margarines, for example, are rich in polyunsaturated fatty acids and low in cholesterol.

Vegetable oils used in margarine include *sunflower, coconut, palm* and *soya bean*. An increasing amount of *rape seed* is used in Europe due mainly to EC subsidies. The bright yellow crop yields about 35–40 per cent oil from its seeds. Early varieties, however, contained a fatty acid called **erucic acid** (22 carbon atoms), the consumption of which is strongly and positively correlated with the incidence of heart disease.

The other constituents of margarine are skimmed milk, whey, salt, vitamins A and D, emulsifiers, and colouring. The skimmed milk is cultured with a starter culture of lactic acid bacteria which acts on the lactose in the milk to produce lactic acid and a number of valuable flavour compounds, including diacetyl. This latter compound has a contributing effect to the overall butter flavour produced by the starter culture and helped by using whey.

By law the maximum water content of margarine must be 16 per cent. Colouring is supplied by annatto or curcumin.

Processing

- The blend of oils and fats is mixed with the cultured skimmed milk preparation in a large vessel with rotating blades
- Other constituents are added and an emulsion is formed. **Lecithin** is now commonly used as an emulsifying agent but **glyceryl monostearate (GMS)** can also be used
- The texture of the margarine is developed in a **votator** which is a closed cylinder within another cylinder. The votator, a type of heat exchanger, cools and works the margarine mix to develop the texture
- The product is then packed, often in plastic tubs as a barrier is required against light, air, moisture and micro-organisms

There are two processes used in margarine manufacture which alter the texture and behaviour of the product:

(a) *Hydrogenation* is a hardening process by which hydrogen is passed through unsaturated oils and converts them to saturated oils by adding hydrogen to their structure. The double bonds of unsaturated fatty acids are saturated by this addition of hydrogen in the presence of a nickel catalyst. Formerly all margarine was hardened by this process, but now only part is

Four lane margarine tub-filling machine

subjected to hydrogenation and then blended with unsaturated oils

(b) *Interesterification* is a process by which the creaming ability of margarine can be improved. This process results in the re-arrangement of the fatty acids combined with glycerol that make up the **triglycerides** of the fat or oils. Saturated fatty acids, such as palmitic and stearic acid, are moved from the centre of the triglyceride to the outside. The process is carried out by heating the fatty acids to about 105 °C in the presence of a catalyst (sodium ethoxide). The process improves the plasticity of the fat and reduces any graininess, thus improving creaming properties.

Nutrition

- Butter and margarine are similar in energy value
- Margarine, however, is a reliable source of vitamins A and D, whereas butter is rather variable. By law the vitamins are added to margarine: 900 µg of vitamin A and 8 µg of vitamin D per 100 g of margarine
- Growth in sales of soft-spread margarines has been considerable in the light of the debate regarding saturated fats. However, some hard margarines are saturated and correspondingly low in polyunsaturated fatty acids; the latter have been shown to have definite health advantages (see Chapter 2)
- Low fat spread margarines are very carefully emulsified products with as much as 57 per cent water.

Use in cooking

The composition of margarines affects their use in cooking:

(a) For use with bread, as a butter substitute, the more expensive soft-spread margarines are more suitable

(b) Margarine can be substituted for butter in many ways, but perhaps does not give such a rich flavour

(c) Flaky pastry is best made with conventional hard margarines

Cooking fats, shortenings and oils

Composition

The composition of this range of products is totally fat and not an emulsion like butter or margarine. Originally lard was used for most applications but it has limited creaming properties and plasticity. A vast range of products is now available to suit all applications. Like margarine, they are manufactured by the processes of **hydrogenation** and **interesterification**.

(a) *Cooking fats* generally are hard fats that have been improved by blending, interesterification, and by whipping in inert gas. This latter process aids creaming and the whipping properties of the fat.

(b) *Shortenings* are fats containing emulsifying agents such as glyceryl monostearate. They prevent the development of a **gluten** network in cake or biscuit making, resulting in a 'short' texture. Some 'superglycerinated shortenings' or high ratio fats can emulsify a larger quantity of water than others. Their use in cake making allows for the addition of more water and sugar, producing a longer lasting and sweeter cake.

(c) *Cooking oils* are becoming increasingly popular, particularly for frying. They can be single oils such as sunflower oil or blends of many types. Popular ones include olive, soya bean, corn, and groundnut oil.

Processing

Most cooking fats, shortenings and oils, in general terms, are of vegetable origin. In reality, many of the raw materials are found in seeds and nuts together with valuable sources of proteins. The oils are extracted either by *pressure expression* or by using *solvents*:

- Best quality oils, such as olive oil, are produced by pressing the peeled and stoned fruit at room temperature
- Some products, however, have to be heated to between 70 and 110 °C, which causes the cells to burst easily to release the oil
- Considerable pressure has to be exerted in a screw press to remove the oil from some seeds

- Any remaining oil in the seed residue is usually removed by **solvent extraction**
- Modern continuous *solvent extraction* systems usually use petroleum ether to extract oil from flaked or ground seeds:

  (a) Fresh oil is added at the end of the process, which works in a counter-current manner to ensure maximum extraction of oil

  (b) The oil is distilled off to leave a crude oil that has many impurities such as pigment, waxes and odorous compounds. It is usually refined in three stages

  (c) Free fatty acids are neutralized with caustic soda to make a soap that is washed away with warm water

  (d) The oil is blended to remove colours using **Fuller's earth** or charcoal

  (e) Steam is then passed through the oil to deodorize it

  (f) Hydrogenation and blending then takes place, often under computer control, to produce a range of oils and fats.

Nutrition

- These products are high in energy but contain no other nutritionally valuable substances such as vitamins A and D
- The oils rich in polyunsaturated fatty acids are recommended in place of hard fats (see Chapter 2)

Use in cooking

(a) *Cooking fats* are best suited for pastry making – the lighter whipped varieties give a quicker result. Hard fats, however, must be used for rubbed-in flaky and puff pastry

(b) *Cooking oils* are chosen when their properties of little or no flavour and a high smoke-point are useful. At the smoke-point (temperature) oils starts to breakdown into smokey fumes and produce unpleasant odours. After each use, the smoke-point of an oil becomes lower. Decomposition during deep-fat frying is reduced by using a tall, narrow heating vessel. There was concern some years ago that continual use of a cooking oil caused the gradual build-up of *carcinogenic substances* in the oil, but this has not been proven conclusively. However, *rancidity* is a problem in many products of this type. The process of rancidity is accelerated by light (UV light) and by metals, particularly copper and iron. For this reason, oils should be stored in glass or plastic containers and not exposed to light. Light from fluorescent tubes contains a high enough level of UV light to trigger rancidity

(c) *Vegetable oils* are being used increasingly as substitutes for solid fats in pastries, biscuits and sauces.

Meat and meat products

Meat

Composition

Meat is not merely the flesh of animals, but results from a series of complex biochemical reactions. Most forms of meat originate from muscular tissue in the animal. The muscle fibres, in simple terms *myosin* and *actin*, combine together after death to form a rigid substance called *actomyosin*. This rigidity in an animal is known as **rigor mortis**, and lasts for some time before the animal's carcass becomes soft again. During this period the original sugar reserve in the muscle is used up and converted into *lactic acid*.

Glycogen is a **polysaccharide** that is broken down in the muscle to produce glucose, a source of energy. Well rested animals have ample glycogen in their muscles, but excited or frightened animals have lower levels. Lactic acid produced from this glycogen lowers the pH of the muscle *post mortem*, usually to about pH 5.6. Lower levels of glycogen result in a higher pH, which produces darker meat that is more susceptible to microbiological decay.

When *rigor mortis* has disappeared, the animal carcass becomes pliable again and the meat becomes more tender and juicier. The meat then passes through a process of conditioning, or **ageing**. Game requires a longer period of ageing and is therefore 'hung' for several days or weeks. During this period enzymes in the meat attack the protein molecules and break them down into simpler, smaller molecules. Amino acids form proteins, and fatty acids from fats are produced and these contribute to the flavour of meat. Connective tissue within the meat is not attacked by enzymes, however, so meat rich in connective tissue will be tough or stringy.

The colour of meat results from the pigment **myoglobin**, which is similar to the blood pigment **haemoglobin**. Both these red pigments combine with oxygen in the normal functioning of the

98

muscle. However, the public prefers joints of meat to be bright red in colour rather than dull or blueish red. The myoglobin must therefore be in its oxygenated form (*oxymyoglobin*) and the packaging or storage of meat cuts must be such as to allow the retention of oxymyoglobin. Lower levels of oxygen result in the production of blueish-red myoglobin and sometimes in a brown pigment (as in cooked meat) called **metmyoglobin**. As mentioned earlier, if the pH of meat is higher than normal it is usually a darker colour. Meat at a pH higher than 5.6 has a greater water-holding capacity, so the protein becomes swollen with extra moisture. This prevents oxygen diffusing far enough into the meat cut to produce ample oxymyoglobin and so myoglobin with its blueish-red colour tends to dominate.

The *muscle fibres* that make up meat are arranged in bundles surrounded by *connective tissue* which give the appearance of grain in the meat. Generally meat is carved across the grain as chewing along the grain is much easier.

Connective tissue is always found in meat but in some cuts it results in a tough product that needs long cooking to make it edible. Three proteins make up connective tissue:

(a) *Collagen* is the most common and makes up the sheaths surrounding the muscle bundles. Levels of this protein tend to be higher in active muscles. On heating with water, collagen becomes gelatine

(b) *Elastin* is the protein component of blood vessel walls and ligaments, and often has a yellowish colour

(c) The third connective tissue is *reticulin* which is rather fibrous, occurring in the spaces between muscle bundles

The texture of meat is probably more important from an eating point of view than flavour. Connective tissue, both the amount and type, affects texture: coarse muscle fibres produce tougher meat. Young animals have fine, small diameter muscle fibres which become larger as the animal ages. Young animals produce the most tender meat, such as veal. Little-used muscles, similarly, produce tender meat – the white breast meat of chicken is more tender than the meat from the continuously-used muscle of the legs.

The best meat has fat, known as *marbling fat*, running through the muscle, breaking up the muscle bundles. Its effect can be likened to the use of shortening in biscuit and cake mixes. This

Cuts of beef: (top) fillet chain, joint; (left to right) chateaubriand, fillet steak, tournedos, tail, tail cut for stroganoff

fat, more unsaturated than the fat found around the joint, melts during cooking keeping the meat moist and succulent.

Beef

Beef is still the most popular red meat and a number of breeds of cattle are bred exclusively for meat. Young male animals give the best meat, particularly from breeds such as Hereford, Shorthorn, Aberdeen Angus and recently Charolais. Older animals tend to produce tougher meat only suitable for canned meat products.

Veal

The texture of veal can be likened to that of chicken, and is at its best when coming from three month old male calves. In other countries 'veal' often refers to young beef and differs little from beef except for being slightly lighter in colour.

Lamb

Lamb is generally tender meat from young sheep, older sheep producing *mutton*. Sheep used to living on hillsides tend to produce tougher meat than animals reared on plains.

Pork

Breeding programmes have produced leaner varieties of pig, that are characterized by the long backed Landrace and Large White breeds.

Pork production can present problems: for example a condition known as **PSE (pale soft exudate)** of muscle can occur in which, post-mortem, the pH of the muscle falls too rapidly and the meat loses some of its water-holding capacity, resulting in the loss of fluid. This is a problem associated with certain breeds and not others.

Processing

- The processing of meat and meat products begins with the *slaughter* of the animal. Modern methods involve the electrical stunning of pigs and the use of captive-bolt pistols for cattle

- The animals must then be thoroughly *bled* by hanging for about 10 minutes as blood is an ideal medium for the growth of putrefacture micro-organisms

- To remove the hair from the carcass, the skin is scalded, usually at an ideal temperature of between 60 and 63 °C for pigs. The hair is removed by a process called *scudding* to leave a whitish skin. Some tough hairs may be singed

- The following parts of the carcass are then removed as cleanly as possible: intestines, stomach, liver, heart, lungs, gut fat, genito-urinary organs, but not the kidneys. Many of these parts of the carcass find uses in the food and pharmaceutical industries

- *Conditioning* (ageing) takes place usually at around 5 °C to control enzymic changes and minimize microbial growth. Procedures vary for different types of meat but the same basic principles apply

- *Meat cutting* is a very skilled operation resulting in loss of value on some cuts and a considerable increase in value of others. The forequarter of beef depreciates, whereas the hindquarter appreciates in price compared with the carcass price. Although there are considerable variations in types of cut, generally the anatomy of the animal is followed to produce the various 'joints'. Long, hard bones are usually removed whole, whilst smaller ones are removed to allow rolling, for example in rolled ribs and loins.

Nutrition

- Meat is a good source of protein, although micro-organisms and some plants produce more protein at a lower price. The proteins of meat are, however, of high biological value with all the **essential amino acids**

- A good range of B vitamins is contained in meat, especially in pork, as well as minerals particularly iron

- The fat content of meat varies according to cut. Depot-type fat, for example from around the kidneys and under the skin, is saturated, whereas fat contained within the meat itself is more unsaturated. Even lean meat can contain about 25 per cent fat

- In 1986 the 'mad-cow disease', *bovine spongiform encephalopathy (BSE)* was identified. Originally thought to be caused by a virus, it is now believed to result from rogue proteins, called *prions*, entering the animal's brain. Concern has been expressed that the disease may be passed to humans eating infected animals. There is no evidence to date to support this

Effect of cooking

Meat is cooked to make it safe to eat by destroying micro-organisms, toxins and parasites. Cooked meat is easier to chew, unless over-cooked, and easier to digest. Cooking also produces the appetising flavour of cooked meat.

During cooking the nutritional value of meat decreases. Heat destroys some of the vitamins, particularly **thiamine** and, to a lesser extent, **folic acid**, **pyridoxine** and **pantothenic acid**. Nutrients are also lost in the meat juices, but these can be regained by using the juices in gravy. Heat also denatures the proteins which can be beneficial as it makes the proteins more readily digestible, giving an apparent increase in nutritive value.

The nature of meat *flavour* has received considerable attention, particularly in light of the need to produce meat flavours for use with meat substitutes such as **textured vegetable protein (TVP)**. During cooking fats melt and mix with water-soluble substances. Complex reactions occur as a result of this between sugars, minerals, enzymes, free amino acids and fatty acids. Browning reactions occur, particularly the **Maillard reaction**, between sugars and amino acids or proteins. The Maillard reaction is a series of complex reactions leading ultimately to the production of brown pigments and many other compounds, both of which contribute to the flavour.

Re-heated meat has a different flavour from freshly cooked meat. This is due to changes in the fat, mainly the development of a certain amount of **rancidity**. This rancidity is triggered by the release of iron in the initial cooking from the red pigments myoglobin, and, to a lesser extent, from haemoglobin in any remaining blood.

Changes in the *texture* of meat during cooking are equally as important as flavour development. Meat should be tender and not tough, juicy and not dry. Fluid loss should be minimized and the conversion of *collagen* to soluble *gelatine* will then be encouraged. Connective tissues like collagen affect the eating quality of the meat. Tougher cuts need longer cooking at lower temperatures to ensure a good conversion of collagen to gelatine.

During cooking:

(a) the muscle fibres contract and then start to uncoil as they begin to denature. Cross-linking of muscle fibres readily occurs to make quite large pieces of meat which are held firmly together

(b) the colour of meat changes due to the inability of the myoglobin to hold oxygen at a temperature much above 70 °C; it thus loses its red colour. The protein fraction (globin) of the myoglobin begins to denature and brown metmyoglobin is produced. Rarely cooked meat will of course retain its red colour

Cooking by dry heat, such as grilling, causes rapid evaporation of juices from the surface of the meat. Muscle fibres contract and may become hard. However, contrary to popular belief, it is impossible to seal the surface completely by any intense heating. Flavour development is due to the browning reactions. Cooking by moist heat, such as stewing, prevents the meat drying out but flavour may be lost in the juices produced. Connective tissue is broken down gradually by this method of cooking to produce a very tender dish.

Meat products

Regulations

Meat products are subject to complex regulations, the *Meat Products and Spreadable Fish Products Regulations 1984*. In the Regulations 'meat' is defined as the flesh, including fat, and the skin, rind, gristle and sinew in amounts naturally associated with the flesh used, of any animal or bird which is normally used for human consumption and, in addition, certain parts of the carcass. These parts include the diaphragm, head meat, heart, kidney, liver, pancreas, tail meat, thymus and tongue.

The name of a particular product dictates its minimum meat content:

- A *burger* must have a meat content of not less than 80 per cent of the food and a lean meat content of at least 65 per cent of the meat
- A *hamburger* must be made of beef, pork or a mixture of both at these percentages
- *Chopped meats* must have a meat content of not less than 90 per cent of the food, again with a lean meat content of 65 per cent of the meat
- *Luncheon meat* must have a meat content of not less than 80 per cent and 65 per cent of this must be lean meat

The Regulations for *meat pies* are more complicated:

- If the pie is cooked, it must have a meat content of not less than 25 per cent
- However, if the weight of the product is less than 100 g and not more than 200 g, then it must have a meat content of not less than 21 per cent
- Pies weighing less than 100 g must have a meat content of not less than 19 per cent
- If the product is uncooked, it must have a meat content of not less than 21 per cent
- In a similar way to the cooked version, a product between 100 and 200 g must contain 18 per cent of meat and below 100 g not less than 16 per cent

In all cases the lean meat content must be at least 50 per cent of the required meat content.

In the case of *sausages, chipolatas* and *sausage meat*, pork products must contain at least 65 per cent meat with at least 50 per cent of this being lean meat. In all other cases a meat content of 50 per cent, of which 50 per cent is lean meat, is the normal requirement.

Pastes and *pâtés* must have a meat content of not less than 70 per cent, of which 50 per cent must be lean in the case of pâtés, but at least 65 per cent in the case of pastes.

The Regulations are comprehensive and include such aspects as the declaration of added water to meat products. Like all regulations they are used to protect the consumer and to avoid malpractice. As a consequence, although penalties for contravention can be moderate to severe, the adverse publicity can be very damaging to a company.

Sausages

Composition

Sausages of various types have been in the diet of many countries for at least 2000 years. There are three main groups of sausages:

(a) The British-type fresh sausage is the most popular in the UK, representing over 90 per cent of the market

(b) Dried fermented sausages, made from cured meats which are then dried are popular in Europe, for example, salami-type products

(c) Cooked sausages, eg frankfurters and liver sausage, are made from a meat emulsion which is filled into a casing and heated in hot water or steam. Some varieties are also smoked

The composition of the British sausage is variable, but most meet the Regulations as described previously. The composition includes, in addition to the meat, pork rind, seasonings, cereals (generally rusk, but sometimes breadcrumbs), additives and water.

Comminuted meat becomes unfit for consumption very quickly, as an enormous surface area is available for micro-organisms to grow, under ideal conditions for some types. The use of chemical preservatives in sausages is therefore essential to guarantee any shelf-life. **Sulphur dioxide** is used (up to 450 ppm (parts per million)) usually in the form of sodium metabisulphite. Sulphur dioxide is an effective inhibitor of food-poisoning organisms such as *Salmonella*. Some spoilage organisms, although inhibited by sulphur dioxide, do eventually grow, thus producing sourness in the product. In general, therefore, sausages will taste sour before there is a risk of food poisoning.

Polyphosphates (0.1–0.3 per cent) are added to help retain moisture in the sausage by increasing the water-binding capacity of the proteins. Cooking losses are reduced by using polyphosphates, and shrinkage of the product is reduced. Some protein supplements such as casein, soya flour and milk powder help retain moisture and fat during cooking.

Processing

The traditional British sausage is made from meat from various parts of the carcass which is either minced or chopped:

- If mincing is used it is normally performed in two stages, with the other ingredients being added at the second stage. Mincing is sometimes then followed by bowl chopping

- In chopping, the bowl chopper rotates and has rotating blades which chop and mix the product. Preservative, spices, colour, rusk and an emulsion of fat (often made with crushed ice) are added to the meat, together with a spice and herb mix – some sausages being richer in certain herbs than others. An example of a typical spice mix is given below:

pepper	75%
ginger	10%
nutmeg	5%
cinnamon	4%
mace	4%
coriander	2%

Sausage manufacture

- The sausage meat is filled under pressure into skins, which are traditionally cleaned, salted sheep intestines (small intestine). Larger sausages are filled into pigs intestines or artificial skins developed from collagen
- After linking the sausages are cooled, wrapped and stored under chill conditions

Nutrition

Sausages are high energy foods because of their high fat content. Raw pork sausage can have an energy value of over 1500 kJ/100 g. However, as fat is lost in cooking this may drop by around 15 per cent after grilling or frying. The development of low fat sausages by incorporating more starch binder has reduced this energy level and also reduced the saturated fat content.

Effect of cooking

During cooking sausages contract and lose water and fat. Poorly made sausages will lose a greater weight than well emulsified and bound products. An attractive browning of the surface is essential and this is sometimes encouraged by the addition of glucose. The glucose undergoes the Maillard browning reaction by reacting with the proteins present. It also serves to lessen the saltiness of the sausage mix.

Cured meats

Composition

In cured meats the meat is combined with salt, sodium or potassium nitrate and/or nitrite, and often sugar and seasoning. Many products are also smoked. *Salt* functions as a preservative as well as giving flavour. The *curing salts* similarly have preservative properties but also produce the characteristic pinkish colour of cured products. In curing brines salt-tolerant bacteria are active and break down the curing salts (sodium/potassium nitrite) to nitrite which is then converted to nitrogen II oxide (nitric oxide). This latter compound combines with the red myoglobin in the pork to produce a pink pigment.

The bulk of curing is confined to bacon and hams. Current thinking about salt levels and curing salts has led to milder cures using considerably lower amounts of each.

Processing (Bacon)

The *Wiltshire cure* is the basis of most bacon curing procedures:

- After slaughter and draining brine is pumped into the sides of pork at about 5 °C They are then left covered in this solution in large tanks. The brine is usually about 25–30 per cent salt and 2.5–4 per cent potassium/sodium nitrate.

- After about five days the brine is drained away and the sides are matured for about two weeks

- During this maturation period the typical flavour develops to produce 'green bacon'. This bacon is preferred by some people, but others prefer a smoked product. Smoking not only adds flavour but coats the bacon in phenolic substances which act as preservatives.

Injected sides of bacon being stacked in curing tanks to be covered with brine for several days

Modern bacon processing is carried out rapidly by slicing the bacon into ready-to-eat slices and curing the slices:

(a) The individual slices are passed for up to 15 minutes through a weaker brine

(b) Maturation can be completed in a few hours

(c) The slices may be smoked by the use of 'liquid smokes' sprayed on by a development from car-spraying technology. The slices are charged electrostatically so that the liquid smoke droplets cling to the surface in an even layer

Nutrition

- The level of *salt* (sodium) in cured products is higher than most foods and frequently exceeds 1 000 mg/100 g

- Concern has been expressed (see Chapter 2) that *nitrites* can combine with constituents of bacon (secondary amines) to produce carcinogenic substances called **nitrosamines**. These substances have been implicated in cancer of the throat. However, the chances of this occurring are very remote. Also, recent studies have shown that ascorbic acid (vitamin C) reduces the formation of nitrosamines

- The *phenolic compounds* of smoke have also been implicated as carcinogens and also may cause allergic reactions in some people

Effect of cooking

Cured foods are not always thoroughly cooked before eating. However, the curing salts, salt and smoking prevent the growth of harmful bacteria. The deadly organism *Clostridium botulinum* which produces a very poisonous toxin is inhibited by this combination of salts.

Traditional smoking processes are, in effect, a form of slow cooking, and so smoked products need only to be warmed before eating.

Poultry products

Composition

Poultry products, once a luxury item, now account for over 35 per cent of all meat sales. There has been an increase in sales of chilled products, particularly portions and processed products such as 'chicken triangles', schnitzels and other novelty products. Many of these products are covered in batter or crumb. The more recent Japanese-style crumb, which is larger and crisper, has been responsible for increased consumer appeal for these products.

The composition of poultry products varies from *white meat* types such as chicken containing 25–30 per cent protein and only 5 per cent fat to *darker meat* products such as duck, containing 11 per cent protein and 43 per cent fat. These figures will obviously vary enormously depending on the method of preparation and subsequent cooking. Little used muscle, such as breast muscle, is light in colour, almost white. Continuously used muscle, such as leg muscle, is well supplied with blood and myoglobin, being darker in colour and tougher.

Poultry processing

Processing

There have been significant advances in bird handling procedures, which are humane and lead to a higher quality product:

- Poultry must be *stunned* prior to slaughter (this is mandatory in the UK). A water bath stunner, with an alternating current of up to 100 V passed through, is usually used
- After being *bled* the birds are immersed in scald tanks of hot water to facilitate *feather removal*. Large scale plants now operate automatic defeathering and evisceration systems
- After evisceration it is important to *chill* the birds rapidly to prevent bacterial growth, particularly food-poisoning organisms such as salmonella. Cold water baths are used for chilling with a counter-flow system. Concern has been expressed that uric acid from the birds can lower the pH of the bath, facilitating the growth of salmonella. The addition of sodium bicarbonate to raise the pH to about pH 8 has therefore been recommended. The addition of chlorine (50 ppm) is also carried out to control bacteria

The production of a range of products from chicken and especially from turkey has led to the development of specialized processing equipment. Although cutting by hand is still carried out in many processing units there are machines to do this at quite a rapid rate. However, machines of this type demand a standard bird of certain dimensions.

Mechanically-recovered poultry meat (MRM) is widely used in poultry products which are of increasing variety. The basic principle involved is to separate the flesh from the bones and gristle. Simple equipment involves pressing the meat through a perforated cylinder leaving the bones behind. The meat is recombined, often with the aid of polyphosphates to increase the water holding capacity.

Nutrition

There has been a growth in poultry consumption partly because of the health aspects associated with red meats and saturated fats. Poultry meat contains *less fat* and a higher content of *unsaturated fatty acids* than red meats. The exceptions to this are duck and goose. White meat contains *less iron* than red meat as there is less myoglobin.

Effect of cooking

It is essential to defrost frozen poultry fully before cooking lest insufficient heat penetrates the bird to destroy pathogenic organisms such as salmonella.

Dark meat is generally less tender after cooking but usually more juicy. White meat is generally much drier and with less flavour. Intense rearing of chickens does not allow development of much darker meat and therefore a lower level of flavour is found in these birds. The other extreme is shown in game birds with dark, tough meat which requires hanging for some time to achieve an acceptable level of tenderness.

Meat substitutes

Composition

A number of products have been produced as substitutes for meat which are protein-based. These products are also known as *novel proteins* or *meat analogues*. The proteins are produced in a number of ways from plants or micro-organisms. The most successful so far has been soya protein which can be produced in various forms and textures, generally known as **textured vegetable protein (TVP)**. Fungal protein, which can be spun and textured, has been produced by continuous fermentation of waste carbohydrates. This is commercially sold as 'Quorn'.

Technology from the textile industry has been employed to produce some of the products, along with current advances in extrusion cooking.

Processing

In producing *TVP* from soya beans the first stage is to produce soya flour, which is defatted. This flour is mixed with water, heated under pressure and extruded to produce a sponge-like consistency. The product is also dried during extrusion and can be cut into cubes or ground so that a range of 'meat-like' products can then be produced. Colour and flavour must also be added to the product during extrusion. The development of meat flavours has taken many years of research and is still under development, the retention of flavour within the TVP still being a major problem.

Spun soya is a second-generation product with a better meat-like texture. The textile industry has been able to produce very fine soya fibres by a special extrusion process. These fibres are pressed into blocks with flavour and colour added. Recent advances include the addition of some fat between the fibres to give the impression of 'marbling' found naturally in meat. Similar processes have been developed for *fungal proteins*, these having the advantage of growing in filaments.

Nutrition

- Vegetable protein products are generally nutritionally close to meat in having a good range of amino acids. However, often there is one or more essential amino acids missing or found in low concentrations. This means that these products may need fortification with some amino acids if they are to be used on a continuous basis instead of meat. Soya protein, for example, is deficient in the essential amino acid *methionine*
- Other problems arise such as a bean-like flavour which has to be removed
- For some people soya causes excessive flatulence and sometimes stomach upsets. This is because of the presence of certain carbohydrates, *oligosaccharides* (for example, raffinose)
- The amino acid composition of fungal protein can be better balanced than soya. However, careful testing of products is necessary to ensure there is no risk of toxicity

Effect of cooking

Products such as TVP would probably have been more successful commercially if they had not been compared with meat. Although, technology has made great advances, the true meat texture and flavour still cannot be totally reproduced.

TVP should be rehydrated sparingly during cooking so as to retain as firm a texture as possible. Rehydrating TVP in gravies or sauces, such as curry, enhances the flavour of the product and makes it more like meat. Particular success has been achieved in meat pies.

Fish

Oily and white fish

Composition

Many of the aspects of meat science apply to fish. However, since fish have to be caught, their supply of glycogen is quickly used up, giving little lactic acid after death. As a result, most fish species spoil rapidly.

Hundreds of fish species are used as food, each varying in composition and processing behaviour. **Pelagic** fish live in the middle and upper layers of the sea and include fatty fish such as herring and mackerel, containing up to 20 per cent fat. **Demersal** fish are bottom feeders and include flat fish, such as plaice and sole, and 'normal' fish, particularly haddock and cod. **Freshwater** fish include salmon, trout and in some areas pike, perch and carp.

A number of factors influence fish composition and quality. Poor fishing grounds that are low in food, lead to a lower fat content in fatty fish. Similarly, fish in spawning are of poorer quality. As the resultant pH of fish is 6.5 or higher, rapid bacterial deterioration is inevitable. A nitrogen-containing compound, *trimethylamine oxide*, is broken down by bacteria into **trimethylamine**, producing the characteristic bad fish smell. In some fish ammonia is also produced.

Processing

As fish rapidly spoils it must be either iced or frozen as soon as possible.

- *Ice* is an ideal medium for chilling fish. Melting ice is actually better for chilling as it absorbs latent heat from the fish to form water. In addition the water produced washes the fish. **Icing** must be carried out thoroughly to extend the shelf-life for several days

- *Freezing* fish down to about −30 °C will extend the shelf-life of most fish to several months. Frozen fish are often dipped in water to give a 'glaze' which prevents dehydration during

Herring on ice

storage. Filleted fish are often frozen in blocks in a vertical plate-freezer. The blocks may subsequently be sawn up to make fish fingers

Fish is preserved by a range of traditional methods, popular throughout the world, particularly *salting, marinating, drying* (often sun drying), *smoking* and *canning*. Smoking can be carried out by cold or hot methods, the latter cooks and preserves the product. Only fatty fish can be successfully canned as the flesh of white fish tends to break up during the heat processing.

Nutrition

- Fish is a highly nutritious food, rich in high quality protein
- The fatty acids in fish are unsaturated and in some fatty fish very unsaturated such as the beneficial omega 3 (or n–3) fatty acids
- There is also an abundance of *essential fatty acids* in fish, particularly *linoleic acid*. These are essential for children, and there is evidence that they may work against the effects of

saturated fats and protect against heart disease
- Fatty fish are rich sources of vitamins A and D. Herrings are one of the richest sources of vitamin D
- It is now thought beneficial for at least three meals a week to be of fish

Effect of cooking

(a) Fish muscle is in the form of blocks, with little connective tissue. On cooking these blocks may separate and collapse – this is clearly shown in haddock
(b) As fish protein readily coagulates and contracts to eliminate water, over-cooking can quickly lead to a rubbery texture
(c) Frozen fish, which has been frozen too slowly, will lose considerable water as 'drip' on thawing, and will cook to give a dry, tough product
(d) The high fat content of fatty fish tends to help keep them from drying out during cooking.

Shellfish

Composition

There are many different varieties of shellfish falling within the category of either molluscs or crustaceans. The most common molluscs (cockles, mussels and winkles) are unusual among fish products in having a high iron content. Most edible crustaceans are ten-legged and include crabs, lobsters, prawns and shrimps. The majority of these spoil rapidly through enzyme or microbial activity. Ammoniacal smells are indicative of spoilage.

Shellfish, because of their feeding habits, can become contaminated with disease or food poisoning organisms.

Processing

- Ideally crustaceans should be kept alive until they are needed
- Boiling is usually carried out to inactivate enzymes which can cause rapid spoilage. Sometimes this is performed at sea by using salt water then chilling in ice, freezing or canning. Generally only white crab meat is canned. Shrimps and prawns are usually boiled immediately to inactivate their enzymes,

found mainly in the head section. This boiling usually brings out the pink colour compared with a somewhat brown live animal. Prawns are usually frozen into blocks

- Although many molluscs are eaten fresh, they can be frozen, canned or smoked. Oysters and mussels can be marinated, usually in vinegar.

Nutrition

Crustaceans and molluscs are good protein sources, but are sometimes oddities in their nutritional contributions. As mentioned above, shellfish can be very high in iron, and some molluscs such as oysters can be high in vitamin C. The infrequency of eating these foods, however, makes them of little consequence.

Effect of cooking

(a) Overcooking can lead to a tough, dry texture in some products

(b) Molluscs should be adequately cooked to destroy any microorganisms present

Fruit and fruit products

Fruit

Composition

Officially, *fruit* is the seed-bearing part of a plant. However, some plant parts that do not come under this botanical definition are called fruits, for example rhubarb which is a stem.

Often comprising at least 90 per cent water, fruit also contains a wide range of **pigments**, acids, flavours, carbohydrates and enzymes. The green pigment **chlorophyll** is found universally in unripe fruit and when this breaks down in **ripening** new pigments are synthesized or existing ones exposed. **Lycopene**, for example, is synthesized in ripening tomatoes (really a fruit) and in other fruits similar **carotenoids** are exposed.

In most fruits *citric* or **malic acid** dominates, although different acids may be more important, such as *tartaric acid* in ripe grapes. Fruit flavours, particularly from citrus fruit, are generally known as **essential oils**, from the word 'essence' in the sense of perfume.

Fruits can be divided into two groups according to their respiratory behaviour after harvesting:

(a) **Climacteric** – in these fruit, often fleshy or tropical, the respiration rate falls slightly then rises rapidly to a maximum (the climacteric) within a few days or weeks of harvesting. This increase in respiration rate is associated with changes in the fruit we know as ripening, such as a change in colour, and the development of softness and sweetness

(b) **Non-climacteric** fruit do not show this burst of respiratory activity but respire at a steady rate and gradually ripen, often on the tree

Examples of climacteric fruit are the banana, mango and avocado. Non-climacteric fruit are exemplified by the citrus fruits.

The gas **ethylene** (ethene) acts as a plant hormone in stimulating the rapid rise in respiration rate in climacteric fruit. The gas can be applied to unripe fruit to stimulate ripening. After the climacteric, the fruit, now fully ripe, will rapidly deteriorate until it is inedible.

Processing

Fruits usually have to be stored for some time before consumption or processing. Obviously climacteric fruits are more difficult to store than non-climacteric fruits:

- Various storage systems, generally known as **controlled atmosphere (CA) storage**, are available which slow down the respiration rate and delay the climacteric. The systems commonly used employ reduced temperature and oxygen levels, or increased carbon dioxide levels
- Tropical fruits cannot be chilled below a critical temperature of about 11–14 °C, whereas temperate fruit can be taken much lower
- Oxygen levels cannot be taken too low, or carbon dioxide levels too high, otherwise anaerobic metabolism will occur, resulting in the production of toxic intermediates and in fermentation
- A reduction of atmospheric pressure (**hypobaric storage**) has extended the storage life of a number of fruits, but it is expensive and difficult to perfect
- A recent system, known under the trade-name of 'Pro-long' involves coating the fruit in an edible layer of cellulose and wetting agents. This layer, when dry, slows down the respiration of the fruit and doubles or even triples its storage life

Many fruits are *dried*, particularly by the traditional method of sun drying (where possible!). Some products are *frozen* and many are *canned*. A recent development is canning in juice rather than syrup.

Nutrition

- Although fruits contain some B vitamins and minerals, their main contribution is vitamin C:
  (a) Brightly coloured and darker fruits contain more vitamin C
  (b) Fruits exposed to sunlight also contain more, and obviously tropical products are richer in this vitamin than temperate fruits

- Some fruits like the banana contribute more energy because of their starch and sugar content
- The avocado is exceptional in contributing fat instead of sugar
- Fruits are a main source of dietary fibre, such as cellulose, hemicellulose and pectin

Effects of cooking

Fruit should always be cooked long enough to extract flavours and to intensify them. However, cooking has the following effects on fruit:

(a) Tissues absorb water during cooking, and fruits may split then wilt. This is clearly shown when baking apples

(b) The texture of fruit always softens during cooking with an increase in juiciness ·

(c) Loss of vitamin C can be considerable especially when cooking in large quantities of water. Acidity protects vitamin C but alkalinity in cooking water can accelerate its destruction.

Fruit juices

Composition

Fruit juices are extracted from a wide range of fruits by processes which involved *crushing, centrifugation* and often the use of *enzymes*. Juices contain flavours, colours, acids, sugars and usually vitamins, particularly vitamin C. Two main types of juice are produced, (a) clear, low viscosity juices and (b) cloudy, higher viscosity juices. There are, of course, a large number which are in between these two main types in characteristics.

(a) The quality of **low viscosity juices** is shown in a clear juice free from suspended matter. To achieve this, enzymes, particularly pectolytic enzymes, are used to break down any colloidal or suspended solids

(b) In **high viscosity juices** it is essential to maintain a 'permanent cloud stability' with most of the suspended matter not settling out to the bottom of the juice. Pectic substances should not, therefore, be broken down as they help to maintain this permanent cloudy stability by increasing the viscosity of the juice. Pectolytic enzymes must be inactivated by some means early in processing to maintain the pectic substances.

Processing

Low viscosity juices, such as apple juice, are produced from fruit which is macerated then subjected to pressing to release the juice. The juice is filtered and treated with enzymes, pectolytic enzymes and **amylases**, to aid clarification. The juice is then pasteurized and bottled. Yield can be increased by adding enzymes to the macerated pulp to break down the cellular structure of the fruit to release the juice. Centrifuges are also used to increase yield in addition to pressing out the juice.

A typical **high viscosity juice** is tomato juice. The higher the viscosity the better the quality:

(a) The *cold-break method* of making tomato juice involves crushing tomatoes at room temperature and extracting the juice. Seeds and skins are filtered from the juice. Although the product has a good colour and natural flavour its viscosity is lower than desired because pectolytic enzymes actively break down the pectic substances present

(b) The *hot-break method* is preferred by industry as it gives a higher yield of increased viscosity juice. Tomatoes in this method are heated and macerated at 85 °C to inactivate the enzymes. The penalty for this is a change in flavour and colour. The addition of hydrochloric acid, subsequently neutralized by sodium hydroxide, ensures the complete inactivation of enzymes during the maceration process. This results in the highest viscosity possible in the juice.

Nutrition

The main contribution to the diet from fruit juices is vitamin C, usually around 50 µg/100 g for citrus juices. Blackcurrant juice is the best source of vitamin C, with four times as much as citrus juices.

Storage of juices

Once a bottle of juice has been opened the vitamin C level falls quite rapidly due to *oxidation*. The ascorbic acid can undergo a type of **non-enzymic browning** which tends to dull or give a greyish appearance to some juice, particularly pineapple juice.

Preserves

Composition

Preserves are made from fruit that is generally slightly under-ripe. Preserves include jams, marmalades, jellies (not table jelly which is made from gelatine not pectin), conserves and candied fruit.

Jams and similar products depend on the interaction of **pectin** (from underripe fruit) with *sugar* and *acid* to form a gel. The sugar is thought to act in such a way as to allow a bridging across of the pectin chains to produce a three-dimensional network in the gel. Sufficient sugar must be present to achieve this, usually about 65–68 per cent, comprising the added sugar and sugar present in the fruit. The sugar is partly **inverted** in the presence of the acid in the fruit to make glucose and fructose, known together as **invert sugar**. This invert sugar is sweeter but also has the ability to dissolve readily in water, thus making it impossible for water to be used by the sugar (sucrose) to make crystals on standing. (Canned syrup may show recrystallization of sugar after a period of time.)

The pectin in the fruit used for preserves must be in the right form; the broken down pectin found in overripe fruit is unsuitable. Ideally the pectin should be in the form of **pectinic acids** which exist as long, sometimes branched, chains which link in gel formation.

- *Jellies* are clear preserves with no suspended fruit
- *Jams* contain on average about 40 per cent fresh fruit
- *Marmalades* are generally made from citrus fruits with peel in various widths
- *Conserves* are expensive preserves with a high fruit content, and sometimes nuts or spirits are added
- *Candied fruits* are slices of fruit preserved by soaking in strong sugar solutions. Crystallized fruit and glacé cherries are common examples

Processing

The fruit used in making jams can be either fresh, frozen or in the form of a purée in cans. The traditional jam-making process uses equal amounts of fruit and sugar. Modern processes vary the amount of sugar according to the sugar content of the fruit and its acidity.

The fruit and sugar is boiled, usually at low pressure to prevent heat damage. The process is stopped when the total soluble solids reach about 68 per cent, which is reflected in an elevation of the boiling point of the jam. Often a preparation of pectin is used so the manufacturer does not have to rely on the natural pectin in the jam.

Jam manufacture: cooking vessels

Nutrition

- Preserves provide some vitamin C and some energy from the sugar
- The pectin content has recently come under scrutiny as a possible means of reducing cholesterol in the body (see Chapter 2)

Use in and effects of cooking

(a) Jams are widely used to fill cakes and pastry

(b) Overcooking can lead to browning, particularly if the jam becomes dehydrated

(c) Other uses of preserves include candied fruits (in cakes), marmalades (to fill cakes), jellies (as glazes), and jam, jelly and marmalade as 'toppings' for toast, bread, scones etc

Vegetables and vegetable products

Vegetables

Composition

Like fruit, vegetables contain mainly water plus varying amounts of starch, sugars, vitamins and minerals, although some are rich in protein. The storage *roots* and *tubers* such as potato are richer in starch. *Legumes*, the peas and beans, contain more protein. *Leafy vegetables* contain some vitamins, particularly vitamin C.

Vegetables behave like non-climacteric fruit and do not show a rapid rise in respiration after harvesting. Stored vegetables can last for several months under the right conditions, but those with a large surface area, such as lettuce or some cabbages, suffer from dehydration and wilt. A false climacteric occurs in some root or bulb vegetables when they start to sprout, the extra energy produced by the rise in respiration being used to produce the sprout and subsequent flowers.

Vegetables, unlike fruit, are harvested when immature. The vegetables are then more tender, often sweeter, and richer in vitamins. Older vegetables become stringy or tough due to deposition of **lignin** on their cell walls. Lignin, a carbohydrate derivative, is found in wood. Legumes such as peas convert their sugar to starch with a corresponding loss in sweetness and an increase in toughness.

Processing

Most vegetables respond well to all types of processing. Enzymes naturally present in vegetables can cause changes in processed products, particularly in texture, colour and flavour. The short heat treatment of *blanching* is therefore necessary prior to further processing, particularly *canning* and *freezing*.

Texture changes in processing can indicate a quality loss in some products. For example, canned potatoes must be firm and without any cracks. To achieve a satisfactory product, calcium chloride is added so that the calcium can bridge across the pectin chains in the potato and, after a period of time, firm the texture. Unfortunately, over-use of calcium salts can lead to a soapy flavour.

Like all high water-content products, vegetables should be frozen quickly. *Rapid freezing* leads to small ice crystals which do not withdraw water from the plant cells as much as large ice crystals developing between the cells during freezing. Vegetables containing more starch or protein can be frozen more easily. A common method used for small products such as peas, is **individual-quick-freezing** (IQF), when cold air is blown upwards in a fluidized bed freezer.

Nutrition

- *Root vegetables* are good sources of starch, a polysaccharide which is broken down into glucose as the root is used to produce new plant material when sprouting. An increase in consumption of complex polysaccharides such as starch is recommended by NACNE

- Potatoes, when new, are also a good source of vitamin C, (30 mg/100 g), as are most leafy vegetables. Since potatoes tend to be eaten in quantity in the UK, they have been a main provider of vitamin C in the British diet. However, levels of the vitamin fall rapidly during storage; after about six months less than half the vitamin content will be left and much of this may be lost in poor cooking

- Some vegetables unfortunately contain components which have adverse effects nutritionally:
  (a) Potatoes, particularly green ones, contain alkaloids (alkali-like nitrogen-containing substances) such as *solanine*. High levels of this can be toxic as found in green potatoes and their sprouts
  (b) Some vegetables, such as cassava, sweet potato and kidney beans, contain substances which contain small amounts of cyanide. Enzymes naturally present in the product can release this cyanide during storage or cooking and in undernourished people poisoning may occur
  (c) The cabbage family contains *goitrins* which can interfere with the uptake of iodine by the body, causing thyroid problems in susceptible people.

Effect of cooking

Cooking has the following effects on vegetables:

(a) Harmful substances may be destroyed or bleached from a vegetable

(b) Starchy vegetables become softer and the starch becomes **gelatinized**

(c) Older vegetables containing lignin remain tough or stringy

(d) Pectin is broken down and made more soluble

Colour changes during preparation or cooking are common.

- Potatoes undergo *enzymic browning*, but at a slower rate than apples. Peeled potatoes may be immersed in water, previously boiled to drive out oxygen, with some salt added to prevent the browning

- The loss of chlorophyll in products like runner beans, is important as a dull-greyish cooked product results. Quick cooking by immersing the vegetables in rapidly boiling water helps retain vitamin C. However, making the water alkaline by adding some sodium hydrogen carbonate (bicarbonate) stabilizes the chlorophyll, thus retaining a bright green cooked vegetable. Unfortunately vitamin C is destroyed readily under alkaline conditions, and is stable under acid conditions – exactly the opposite of chlorophyll. However, the loss of vitamin C in runner beans is not significant in a balanced diet, and an attractive green colour improves the look of a meal

Pickles

Composition

In pickles *acid* is used to preserve the product by discouraging the action of enzymes or micro-organisms. Acid can be added, usually as vinegar, or produced in the product by **fermentation**. Many pickles also have a high *salt* level which helps the acid in its preservative action. The acid in pickles (acetic in vinegar or lactic in fermented products) is in an undissociated form. This *undissociated acid* can enter the bacterial cells whereas *ionised acid* is prevented from entering by the electrically charged bacterial cell wall. When inside the bacterium, the acid dissociates and produces ions, including hydrogen ions. The latter lower the pH in the cell and disrupt the bacterium's metabolism by preventing the action of certain enzymes. This results in the death of the cell or prevents further cell division and growth.

Any product can be pickled but the main ones include cauliflower, cucumber, onion and cabbage.

Processing

(a) The *simple method* of pickling involves cooking the vegetable to a soft texture and then soaking it in a weak brine to draw out some of the moisture. This is followed by immersion in vinegar, sometimes sweetened and spiced

(b) In *fermented pickles* a stronger brine is used to prevent the growth of undesirable bacteria, but the brine is weak enough to allow *lactic acid bacteria* to grow. *Sauerkraut*, for example, is a product made by fermenting cabbage. The fresh cabbage is sliced and salted. Water is then added to achieve a salt level of around 2.25 per cent. The cabbage is left at a temperature of about 20 °C to allow the lactic acid bacteria to grow and produce lactic acid. The fermentation starts with *Leuconostoc mesenteroides* which produces lactic acid up to a volume of about 1 per cent. When this organism dies out *Lactobacillus plantarum* takes over to bring the volume of lactic acid in the cabbage up to about 2 per cent. Sometimes as many as four organisms may produce the lactic acid, all of which are naturally present and not added in a starter culture

Some pickles are pasteurized after bottling to prevent the growth of yeast.

Nutrition

Acid in pickles helps to retain vitamin C in products which are high in the vitamin. However, long storage and processing can deplete levels considerably. The products are consumed in small quantities so this is of little significance.

Storage of pickles

Generally pickles keep without difficulty. Sweeter or low acid products may become infected with yeasts which sometimes appear as yellow spots on the pickled vegetables. This can cause the product to look unattractive but may actually improve the flavour.

Fermented products tend to suffer more problems as sometimes the wrong bacterial population develops, producing pickles that are soft and often taste 'off'.

Cereals and cereal products

Flour

Composition

Flour is produced by milling wheat and has a variable composition depending on a number of factors. The distribution of nutrients is uneven throughout the wheat grain. Many of the B vitamins are in the outer bran layers, whereas starch is concentrated in the endosperm with vitamin E in the germ (shoot and root). Wholemeal flour will therefore have a different composition of vitamins and fibre from white flour. For example, wholemeal flour contains 0.5 mg of vitamin B_1 per 100 g of flour, whereas white flour has only 0.3 mg/100 g; the figures for niacin are 6.0 mg and 2.0 mg/100 g, respectively. (Source: A. E. Bender, *D A Bender Food Tables* (Oxford University Press, 1986))

In producing flours of different characteristics for baking, a number of different wheat vitamins are blended. Wheat which is rich in **wheat protein** (about 13 per cent) is called *strong* wheat and is used to produce loaves of good volume and texture. *Weak* wheats contains about 8 per cent protein and are used to make cakes and biscuits or French-style bread. Traditionally, wheats from North America have been strong compared with weak European wheat. However, this is now changing with the successful growth of new strong wheat varieties in England such as 'Avalon'. It is also now possible to extract protein from strong wheat and to add this to weaker wheats to improve their bread-making quality.

Processing

The milling of wheat to produce flour is divided into two main operations, the cleaning and conditioning of the grain and the size reduction and separation of the flour.

- *Cleaning and conditioning:*
    (a) Preliminary cleaning of the wheat grain may take several

130

operations to remove stones, straw, seeds of other plants and metal particles. Sieves and separators of various kinds are used. Chaff and straw can be separated using aspirators or winnowers

(b) The grain is usually washed in a moving stream of water and the surplus removed in a 'whizzer', a type of centrifuge

(c) The grain is then conditioned to achieve a final moisture content of 15–17.5 per cent. At this moisture level the starch in the endosperm is friable and will separate easily from the bran

- *Size reduction and separation:* The milling process involves cracking open the grain and then disintegrating the endosperm and separating out parts not required:

   (a) The first stage is the use of *break rolls* which have grooves set at a required angle. One roll rotates 2 times as fast as the other. When a grain passes between the rolls, the slower one holds back the grain while the faster one cuts into and partially crushes it

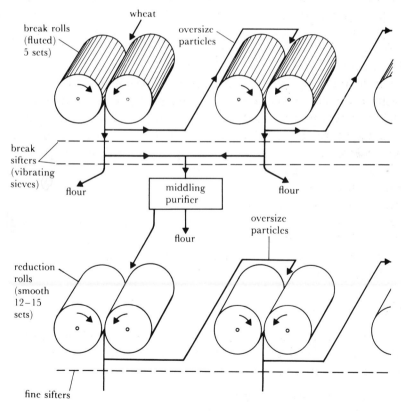

Flour mill

(b) The large particles thus produced may be passed to second break rolls set closer together

(c) After each stage sieves are used to recycle large particles

(d) The next stage uses several reduction rolls which are set very close together and gradually produce a fine flour

In flours with a *low* **extraction rate**, the white endosperm has been removed from the grain and no bran is present. *High extraction rate* flours contain much of the outer layer of the grain, for example wholemeal flour has an extraction rate of 95 per cent compared with 75 per cent for white flour.

Flours must be aged to improve the performance of their protein in bread-making. The process can be accelerated using chemical **improvers** such as potassium bromate, chlorine dioxide and ascorbic acid.

Nutrition

- Flour, and consequently bread, is a good source of *protein, B vitamins* and *carbohydrate*
- The carbohydrate is complex carbohydrate, mainly starch, and so is recommended by the NACNE report
- Wholemeal flour makes a valuable contribution to fibre intake
- *Phytic acid*, however, contained in the fibre may absorb some calcium from the diet, so purified chalk (calcium carbonate) is often added to bread flour
- By law the vitamins **thiamine** and **nicotinic acid**, and iron must be added to white flour

Effect of cooking

Flour is a very useful *thickening agent* and has been used in many products and dishes such as gravy, soups and desserts. The starch in flour **gelatinizes** in the presence of water during cooking. This process is described below:

(a) Since starch has the ability to absorb a large volume of water, when it comes into contact with liquid the starch granules swell and eventually burst

(b) The starch particles that leak out of the granules link across to each other and to intact granules, producing a three-dimensional network

(c) This network contains within it a large amount of entrapped water and, in this way, the product becomes thickened.

(d) On standing, however, the starch network may start to contract and squeeze out more of the water. The process is called **retrogradation** and can be seen sometimes in gravies which have become cold and left for some time

Bread

Composition

The main ingredient of bread is flour, usually from strong wheat, to which is added water, salt, yeast, and often a small amount of fat. When the ingredients are mixed the bread-making process begins as the water is absorbed and hydrates the flour proteins:

(a) The yeast begins to ferment any sugar present, producing carbon dioxide and some alcohol

(b) Naturally occurring enzymes (**amylases**) in the flour start to break down damaged starch from the milling process. This provides a supply of **maltose** which can be fermented by the yeast

(c) Carbon dioxide is produced in the dough which, as a consequence, begins to rise.

The **wheat proteins** play a central role in producing loaves of good volume and texture. Collectively the proteins are known as **gluten** and contain two main proteins *glutenin* and *gliadin*. These protein chains are cross-linked in the gluten and so give it noticeable *elasticity*, which enables the gluten to retain the carbon dioxide as it is produced in fermentation. However, the dough must rise to produce a loaf of the right texture and so the gluten also has to be *extensible*. These features are achieved mainly by **disulphide bridges** (—s—s—) between the protein chains and their interaction with sulphydryl group (—SH). The disulphide bridges give the elasticity to the dough but are reduced back to sulphydryl groups periodically, which then allows the dough to expand.

When the dough has expanded sufficiently during fermentation the baking stage begins:

(a) During baking the yeast is inactivated and some carbon

133

dioxide is driven from the dough. However, due to increase temperature, the gas within the dough expands, thus increasing the volume

(b) The proteins coagulate and the starch is gelatinzied to form the honeycombed structure of bread

(c) The crust browns due to the Maillard reaction between proteins and sugars or dextrins produced in fermentation by break down from starch. The flavour of the crust is due to this reaction and the entrapped alcohol produced during fermentation which is converted in part to esters with some flavour

Bread-making

The *traditional method* of a 'one-sack' dough uses 127 kg (280 lb) flour (the traditional sack), about 2–3 kg salt, 1.5 kg yeast and 70 litres of water (which must be at around 25 °C before being mixed with the flour):

• Fermentation is carried out in a warm place for about 2 hours

• The dough is re-mixed, or 'knocked back', to ensure even fermentation by having constant temperature throughout the dough

• After a further fermentation period of about an hour the dough is weighed into portions and roughly shaped

• After this the dough is rested in the 'first proof' for about 15 minutes, then moulded into shape and placed in tins

• The 'final proof' is carred out in the tins for about 45 minutes during which time the dough rises

• Baking is carried out at around 232–260 °C for 40–50 minutes

• Steam is sometimes injected towards the end of baking to give an attractive glaze to the crust.

A number of **rapid dough processes** have been developed to replace the slow traditional method. The *Chorleywood Process* is very popular and produces consistently good loaves even from weaker flours. The process depends on mechanical development of the dough by thorough mixing: The extended dough is rapidly oxidized by the use of ascorbic acid, which is converted to the oxidizing form, dehydroascorbic acid, in the dough. The disulphide bridges (—s—s—) are thus developed to give the elasticity to the dough. Extra fat, water and yeast are needed. The mixing process corresponds to the slow stretching achieved traditionally by fermentation. However, a more regular texture is obtained in the bread.

Nutrition

Obviously the comments made on flour apply also to bread. It is a main contributor of *protein* to the diet, even though the protein is deficient in the essential amino acid, lysine. Wholemeal breads provide a valuable source of *fibre*.

Staling of bread

The familiar staling of bread occurs as a result of changes in the starch, which was gelatinized during baking. The process of **retrogradation** occurs and the starch loses water which is absorbed, in part, by the protein and also by the crust. The crust as a result becomes soft and leathery and the breadcrumb becomes dried out and firm. Rapid freezing of bread delays staling but the process is accelerated after thawing the loaf. Chilling to around 7 °C accelerates the staling process.

Staling can be reversed by reheating the loaf to the temperature of starch gelatinization, about 60 °C. However, no moisture must be lost during the process otherwise an even tougher, staler product will result.

Cakes and biscuits

Composition

In addition to flour, cakes contain sugar, fats, eggs and often **baking powder**. The latter is used as a raising-agent in a similar way to the use of yeast in bread. The gluten network in bread retains the carbon dioxide produced during fermentation. This network is undesirable in other baked products because of its elasticity. Flours are generally milled for cakes and biscuits from soft wheats, low in protein. Sometimes, by careful sieving the protein is lowered further to produce *high-ratio* flours which are used commercially for cakes.

The high moisture content of cakes makes them eventually susceptible to mould growth. Mould inhibitors are sometimes used commercially such as sodium or calcium propanoates (propinates) and ascorbic acid.

Production of cakes and biscuits

The cake ingredients are mixed with water to produce a batter into which air is whipped. This air is retained with the help of egg and, in some products, by creaming with a suitable fat. The air is supplemented by the production of carbon dioxide from the baking powder which aerates the batter sufficiently. **Baking powders** contain sodium hydrogen carbonate (bicarbonate of soda) and an acid, which is usually slow acting. Traditionally, **cream of tartar** has been used as the acid, but recently acid calcium phosphate and glucono-delta-lactone have been used.

Sodium aluminium sulphate is an acid which is only active at higher oven temperatures and has an advantage over other

Biscuit dough being cut ready for baking

powders which tend to produce gas too early. The aerated texture of the cake becomes fixed as the temperature in the oven rises to denature the protein and gelatinize the starch.

The choice of fat in some batters is made to interfere with gluten development and make the cake more tender, hence the term *'shortening'* for the fat. *Emulsifying agents* are added to allow the addition of more water and more sugar. A sweeter, longer lasting cake is produced as a result. *Sugar* also interferes with gluten production and contributes to a more tender cake. High sugar levels, however, require more mixing.

The dough-like batters for *biscuits* must be kneaded and rolled, as a large amount of fat is used to disrupt the gluten. Any elasticity in the dough would result in shrinkage when the biscuit had been cut to shape. Only a small amount of baking powder is required to produce a minimum of gas, and baking is usually carried out at a higher temperature to produce a crisp product.

Nutrition

- Because of their sugar content, cakes and biscuits have a much higher energy content than bread, sometimes around double at over 2 000 kJ/100 g for biscuits and about 1 500 kJ/100 g for cakes
- Fat levels can be significant in biscuits at about 20–25 per cent. This fat can be mainly saturated

Staling

- Similar processes to those in bread occur during staling
- Cakes, however, often have a higher initial water content, as mentioned above, which delays the effects of staling
- Mould growth can be a problem when cakes are stored in tins

Pasta

Composition

The *Durum-type wheats* used for pasta manufacture have a higher protein content and tougher endosperm than those used in bread and cake making. They tend to produce yellow *semolina* on milling, with a little free starch and large pieces of endosperm protein with some starch granules attached. This semolina requires less water to make a dough and therefore, less drying after forming the pasta.

There is almost an infinite variety of pasta products, some being enriched with egg or soya flour.

Processing

- The semolina is firstly mixed with water and salt. The resultant dough contains about 25 per cent water compared with around 45 per cent in bread dough
- The dough is kneaded and worked often between heavy rollers
- After a short resting period to develop some plasticity the dough is extruded through a die (in a special press) to obtain the right shape and size
- The extruded pasta is cut into lengths and then dried

Drying is the difficult part of the process as the moisture level must be reduced carefully from 25 per cent to 10 per cent. Too rapid drying results in the outer layers drying ahead of the inner layers which causes cracking. Slow drying allows microbial contamination and perhaps subsequent spoilage. Commercial drying times vary from 15–40 hours, with variable drying rates within that time, usually with the fastest rates at the beginning.

Nutrition

- Pasta is low in *fat* and contains around 84 per cent *carbohydrate*
- The *protein* level is higher than many cereals, but may be reduced on boiling
- Wholemeal pastas are being introduced to improve the *fibre* content of the product

Effects of cooking

(a) Pasta should be boiled only enough to absorb sufficient water to make it limp; it should still be slightly resistant to chewing. This is known as *'al dente'*, or 'to the teeth'

(b) After cooking pasta products have a tendency to stick together, which presents a problem in serving. This can be resolved using one of two methods:
    (i) Pouring boiling water over the pasta after cooking helps

remove some of the loose starch that causes the pasta to stick together

(ii) Mixing drained pasta with some vegetable oil coats the gelatinized starch and prevents it causing the pasta to stick together

Breakfast cereals

Composition

Two main types of breakfast cereal are consumed:

(a) For centuries, various grains, particularly oats have been cooked in water. The water softens the product and gelatinizes the starch, making it readily digestible

(b) The second type originated in the last century and involves preparing the grain by pre-cooking so that it can be eaten with cold milk. This pre-cooking gelatinizes the starch in the endosperm which is then dried. On the addition of cold milk, the starch is able to thicken instantly. A wide range of grains has been used in making these breakfast products which have grown in popularity increasingly over the last few years

Processing

Cornflakes

(a) White maize is used to make cornflakes and must be thoroughly cleaned with all bran and germ removed

(b) The endosperm is ground into pieces which are steamed to gelatinize the starch. This may take 2 or 3 hours

(c) To this gelatinized starch additions are made of vitamins, sugar, salt and malt

(d) After cooking the mixture is pressed between smooth rollers using considerable pressure

(e) The flakes produced are soft and must be made crisp by being heated in an oven. The oven process toasts the flakes and gives the typical appearance of the product

Puffed cereals

Many grains have been made into puffed cereals, particularly rice, rye, wheat, maize and barley:

(a) A special puffing gun is used into which the grain is filled

(b) Steam is then injected which increases the pressure and cooks the grains

(c) The sudden release of the pressure generated by the steam causes the grains to expand to double their size. This expansion is due to the rapid expansion of water vapour within the grains and gives a light porous texture to some products

Shredded wheat

(a) Whole wheat is pressure-cooked to gelatinize the starch in its endosperm

(b) The resultant soft grains are shredded by special rollers producing a continuous flow of strands of cooked wheat

(c) These strands are pressed together in layers to form the 'biscuit'

(d) To prevent the strands from coming apart the ends of the biscuit are crimped

(e) Drying is then carried out in an oven at around 250 °C for 20 minutes

Nutrition

- These processed cereals were once considered 'empty calories' as many of the vitamins are destroyed in processing

- However, most products are now fortified with *iron, niacin* and *vitamins* C, B_1, B_2, B_6 and D. In many cases they can supply between a quarter and a third of the recommended daily amounts of these vitamins

- Many products are also recognized as good sources of *fibre*. Some products have around 30 per cent added bran. Fibre levels are often of the order of 12–15 per cent of the product

- *Protein* levels are usually around 10 per cent

- The milk used with the cereals also provides its nutrients. Ideally, semi-skimmed or skimmed milk should be used to reduce the fat intake

Beverages

Soft drinks

Composition

Carbonated soft drinks originated in the 18th century and now represent a very large industry. The drinks are acidified, coloured, sweetened, artificially carbonated and often chemically preserved. The particular flavouring is often the secret to a successful soft drink. The formulation of some of the more popular drinks are closely guarded secrets.

The carbon dioxide, used in carbonating the drinks to give the fizz, was originally produced by the action of a weak acid and bicarbonate of soda. The use of the latter gave rise to names such as 'strawberry cream soda'. *Coca-Cola* originated in Atlanta in 1886 and contained extracts of coca leaves and cola nuts. The **caffeine** content made earlier drinks somewhat addictive.

Processing

- *Water* used for soft drinks must be purified to a high degree. Whilst being free of micro-organisms, it must be free of dissolved metals and organic compounds. Processing involves mixing the drink, then carbonating followed by bottling

- *Sweetening* of drinks has traditionally been carried out using syrup to give the drink about 12 per cent sugar on average. This sweetening gives body to the drink and improves mouth-feel. In 'diet' type drinks sugar is replaced with **saccharin** or, more recently, **aspartame**. The loss of body produced by the sugar is rectified by the addition of a little pectin or carboxymethyl cellulose

- The *colouring* and *flavouring* of the drinks present problems. Although natural products may be in favour with the public, they are difficult to use to produce a standard product. For example, natural flavour extracts undergo changes in the presence of light, acid and on storage. Also they do not often possess pigment of sufficient depth and are usually unstable in acid conditions. On the other hand, synthetic products can be

141

produced to overcome all the problems encountered to produce a cheap consistent product

- *Acidity* is provided by weak acids, particularly citric, malic and tartaric, and phosphoric acid in cola drinks. Dissolved carbon dioxide also provides some acidity
- *Long-term keeping* quality is ensured by the use of preservatives such as sodium benzoate
- *Carbonation* of drinks is carried out under chilled conditions, when carbon dioxide is more soluble than at higher temperatures. Modern techniques involve exposing a film of the drink, passing over a chilled surface, to pressurized jets of carbon dioxide
- *Bottling* plants are geared to high speed production. This has been facilitated by the use of hardened, lightweight glass, and plastic bottles (PET particularly)

Nutrition

- Soft drinks have no nutritional value, except for the 'empty' calories of the sugar used for sweetening
- The rapid growth of 'diet' products has encouraged the use of alternatives to sugar bringing energy values down to a minimum
- Addition of vitamin C has been made to some products

Tea

Composition

Tea is produced from an evergreen tea bush grown mainly in the Far East, particularly China and India. The bush yields leaves suitable for tea-making after 3 years and up to 50 years.

Tannins are an important constituent of tea as they provide colour, astringency and some body. Some essential oils provide the aroma, particularly in expensive teas. The use of enzymes contained within the leaves can lead to the production of black leaves. The enzymes responsible are **polyphenolases**, the enzymes which cause enzymic browning.

Processing

In processing tea, enzymic activity may or may not be employed to produce three main types – black, green and oolong.

- *Black tea* – the following process is carried out to obtain leaves suitable for use:
  (a) The leaves are allowed to wither and soften, encouraged by partial drying
  (b) The withered leaves are passed between rollers which crush the cells and release, amongst other constituents, the enzymes
  (c) Fermentation then occurs as the cells are ruptured. This process can take as long as 5 hours, but the time can be reduced to between 1 and 3 hours at a temperature of 27 °C. The fermentation does not involve yeast or bacteria, but relies on the naturally present enzymes
  (d) Following fermentation the leaves are 'fired'. Firing is a slow drying process, at a temperature of around 93 °C, that brings the moisture content to below 5 per cent
  (e) The tea is then graded, blended and packed

- *Green tea* must not undergo fermentation, so the enzymes are inactivated by steaming the leaves before rolling

- *Oolong tea* is a compromise, being only partially steamed, allowing some enzymic activity to occur

Nutrition

In itself, tea has no nutritional value, but it does still have an effect on the body:

(a) Drinking tea with a meal reduces the absorption of iron that may be found in the food. This is due to the *tannin* content of tea

(b) The stimulating effect of tea comes from **caffeine** – the world's most common drug! In small quantities its stimulating effects improve concentration, gastric secretions and co-ordination. It also stimulates muscles and makes them less susceptible to fatigue. However, excessive consumption of tea can cause heartbeat irregularities, lack of sleep and nervous problems

Brewing tea

Everyone can brew tea!

- The flavour, and also the caffeine, take longer to extract from the leaves during brewing than the colour. Ideally around 3–5 minutes brings out the best quality
- People preferring weaker tea should use less tea but still brew for this time
- Re-using tea leaves leads to the extraction of bitter components which give a more harsh flavour
- Astringency, caused by the tannins, is reduced by the addition of milk. This is due to the fact that tannin binds to milk proteins and therefore cannot produce the same astringent effect in the mouth

Coffee

Composition

Coffee trees originated in Africa and there are now two species of tree that dominate the production of coffee: The original Arabian style coffee comes from *Coffea arabica*, but the milder *Coffea robusta* is easier to grow and quicker to fruit. Most commercial coffees are blends of different varieties from several regions.

The composition of coffee is complex with a large range of compounds including **flavonoids**, **chlorogenic acids**, **nicotinic acids**, and, of course, **caffeine**. The final composition is influenced by the variety of coffee bean, growing conditions, fermentation and roasting.

Processing

The fruit of the coffee tree is referred to as a 'cherry' and contains two coffee beans surrounded by a tough skin. After harvesting it is subjected to the following processes before appearing as the well-known roasted coffee bean:

(a) The cherries are pulped in a special machine to release the beans, which are covered by a layer of mucilage

(b) Natural fermentation of the mucilage occurs to leave the beans

(c) The beans are then dried to a level of about 12 per cent moisture. Some flavour and colour changes occur during this process. (NB Sometimes the cherries are allowed to ferment before the beans are separated out, in order to soften the pulp and facilitate the process)

(d) The beans are transported to a processing unit where they are sorted and graded

(e) Dry shells of the beans are removed by friction

(f) The beans are roasted at about 205 °C for 5 minutes to develop their colour and flavour – browning reactions are responsible for both of these changes. Dark, strongly flavoured coffee will be produced from hotter or longer roastings

Coffee cherry containing two beans

To make *instant coffee* the beans are carefully brewed after grinding. Large percolators are used and the liquid is drawn off and dried. Most coffee is spray dried, but high quality products are freeze-dried.

Decaffeinated coffee is growing rapidly in popularity. The beans are treated before roasting by steaming and then the caffeine is extracted with organic solvents such as chloroform. Liquid carbon dioxide has been used effectively as a solvent by some manufacturers. Concern over the solvent residues has encouraged the use of steam to bring the caffeine to the surface of the bean which is then removed by friction. Some loss in flavour results from this treatment.

Nutrition

Coffee has been investigated many times as a possible health hazard:

- Like tea, coffee contains caffeine with all its associated effects on the body
- There may be connections between excessive coffee consumption and heart disease, high blood pressure and some cancers
- An increase in concentration of fatty substances in the blood has also been suggested

It must be stressed, however, that there is no conclusive evidence to support these theories at the moment.

Brewing coffee

(a) Coffee beans should be kept whole and cool before grinding

(b) Contact with air reduces flavour and can cause some rancidity

(c) Ground coffee rapidly loses volatile flavours and becomes stale

(d) Finely ground coffee is used to make stronger brews

(e) Brewing should be carried out quickly – like tea, overbrewing leads to the extraction of bitter components. It is estimated that, during brewing, a maximum of 20 per cent of the solids of the coffee should be extracted

(f) Boiling water should never be used since it scalds the grinds

Wine

Composition

Wine has been made in various forms for at least 5 000 years. Although many fruits can be used, the grape has been the most popular for wine-making. The composition, quality and drinking characteristics of wine depend on:

the variety of grape

the soil
the climate
the method of production

White wines are produced from the juice of any variety of grape, not just green ones. *Red wines*, however, can only be produced from black grapes, which are pulped and fermented with the skins present. The pigments of the skin (**anthocyanins**) are extracted during the initial fermentation, and the alcohol produced by fermentation accelerates the colour extraction. Tannins are also extracted which give astringency to the wines. *Rosé wines* and light reds are only allowed to ferment in the presence of the skins for a short time, perhaps a day.

The juice of the grape contains sugars which are readily fermented by yeast (*Saccharomyces ellipsoideus*). If all the sugars are fermented a dry wine is produced, sometimes with an alcohol content as high as 14 per cent (in colder countries this may only reach around 9 per cent). Fully ripened grapes contain so much sugar that some may remain to produce a sweeter wine. In very sweet wines such as *Sauterne* the grapes become shrunken and raisin-like due to the action of a mould, *'la pourriture noble'* ('noble rot'). This mould concentrates the sugars and produces glycerol which increases the sweetness of the finished wine.

Ripe grapes contain **tartaric acid**, which may crystallize out of wines on standing as cream of tartar. Red wines and white wines from unripe grapes, for example *Hock* and *Moselle*, contain more **malic acid**. In *semi-sparkling* wines this malic acid is converted by bacteria into lactic acid and carbon dioxide which gives the slight effervescence. Full sparkling wines, like champagne, require special alcohol-tolerant strains of yeast and fermentation is completed in the bottle over a long period.

Wine-making

Wine-making can be carried out with a minimum of equipment, using the yeast naturally present on the grapes, seen as a bloom. However, the product is rather variable and generally not of a commercial quality. Normally the following process is employed:

(a) The grapes are crushed and treated with sulphite to kill the wild yeasts and any bacteria present

(b) An active starter culture of yeast is added and fermentation proceeds at around 22–25 °C

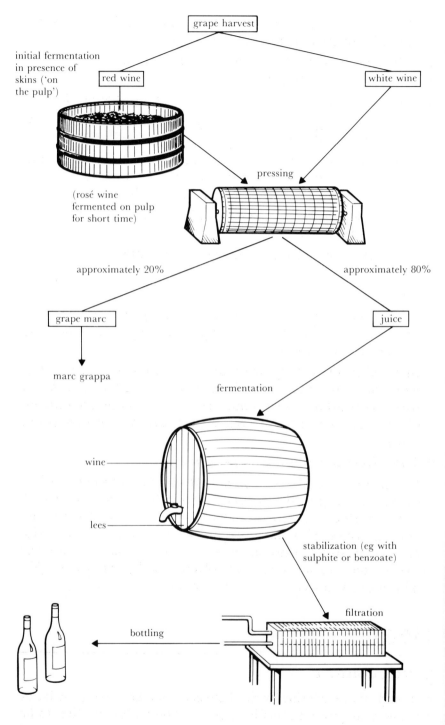

The vinification process

(c) Fermentation stops when all the sugar is exhausted or the yeast is poisoned by high alcohol levels

(d) The yeast falls to the bottom of the vessel and may start to breakdown or autolyse

(e) Racking is carried out to remove the yeast and other debris

(f) Several rackings may be necessary as the wine is aged in tanks or ideally in oak casks

(g) Red wines, because of their tannin content, require many years of maturation (**ageing**) in order to produce a more mellow, better flavoured wine with a good bouquet

(h) After maturing wines are normally filtered and stabilized by adding sulphite or benzoate

(i) Further maturation is carried out after bottling

Nutrition

- In one sense, wines are nutritionally insignificant since they provide the body with only calories, especially sweet wines, and some iron (1 mg/100 g)

- In another sense they are more significant in that slight to moderate consumption has been shown to reduce the risk of coronary heart disease in some people by effecting the removal of fat deposits in the arteries (although this is difficult to prove conclusively)

- The effects of excess are well known

- Health risks from wine also arise from adulteration which is carried out to give the appearance of a better quality wine. A cheaper Hock may be sweetened to give the flavour and characteristics of a more expensive 'auslese'. This was discovered in Austrian and some German wines and was caused by the addition of an industrial solvent, diethylene glycol. (NB This is not anti-freeze which is ethylene glycol)

Beer

Composition

Although there are similarities between wine and beer, the latter is made from *grain*, particularly *barley*. The barley has first to be germinated in the malting process so that **amylases** can break down some of the starch to **maltose**. It is this maltose which is fermented by the yeast, whereas in wine the grapes contain sugars that are readily available.

Although beer has been made for thousands of years, two innovations occurred in the Middle Ages which greatly affected it:

(a) Hops were used to preserve the beer and flavour it with the bitter *humulones* they contain

(b) Lager was produced around this time following the discovery of bottom-fermenting yeasts as opposed to the traditional top-fermenting yeasts of other beers

Further developments have included the use of wheat grains and roasted grains in addition to malt to give body and colour to the beer. Also the head of the beer has been stabilized for a number of years by the addition of a small amount of extract from horse-chestnut.

Beer-making

(a) After preparation of the malt, the malt and cereal grains are mixed and hot water added. The temperature is controlled at around 65 °C to encourage rapid amylase activity. This process is called *mashing* and is carried out in a large vessel called the *mash tun*

(b) After about three hours the liquid is drained off. This liquid, the *wort* (pronounced wirt), is rich in sugars

(c) The wort is boiled in a large copper and then, after some time, the hops are added

(d) The wort is filtered, cooled and a yeast starter added

(e) Fermentation is carried out at about 15 °C. In traditional beers the production of carbon dioxide during fermentation carries the yeast to the top of the vessel. In lager the yeast ferments at the bottom and as a result is easier to separate after fermentation

(f) After fermentation *finings* are added and filtration is carried out to remove yeast cells

(g) The beer is then stored in bulk prior to bottling or in kegs

(h) Traditionally some sugar was added in the bottle or keg to achieve carbonation of the beer. Modern plants have carbon dioxide injection systems to achieve a consistent level of carbonation

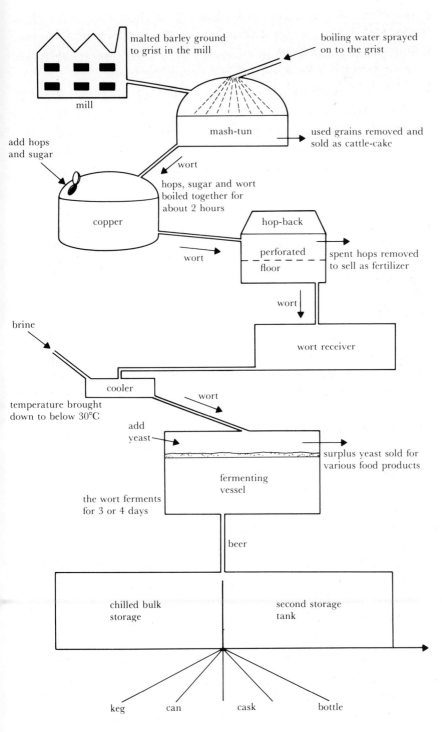

malted barley ground to grist in the mill

boiling water sprayed on to the grist

mill

mash-tun

add hops and sugar

used grains removed and sold as cattle-cake

wort

hops, sugar and wort boiled together for about 2 hours

copper

hop-back

wort

perforated floor

spent hops removed to sell as fertilizer

wort

brine

wort receiver

cooler

temperature brought down to below 30°C

wort

add yeast

surplus yeast sold for various food products

fermenting vessel

the wort ferments for 3 or 4 days

beer

chilled bulk storage

second storage tank

keg   can   cask   bottle

The brewing process

Nutrition

- Beers are good sources of *energy* and some are good foods because of their *carbohydrate* and *protein* contents
- Beers are richer in *minerals* than wines, but lower in *alcohol* at only 2–5 per cent
- Some B group *vitamins* produced by yeast also appear in the finished beer

Sugar confectionery

Chocolate

Composition

Chocolate, produced from *cocoa*, originated in South America, and was brought to Europe by early Spanish explorers. Cocoa beans grow in pods containing around 25 beans. From the beans *chocolate liquor* is prepared, the essential ingredient of chocolate. After removal of the liquor the remaining cake is processed into *cocoa powder*. Weak alkali is often added to improve the solubility and darken the colour. Sugar, vanilla essence and salt are added to produce *drinking-chocolate*.

The chocolate liquor is rich in fat, around 55 per cent, and contains approximately 17 per cent carbohydrate and 11 per cent protein. Around 6 per cent tannin is also found. Chocolate contains two alkaloids, *caffeine* and *theobromine*. Although the caffeine content is small and has little effect, theobromine is a stimulant (diuretic) and can be slightly addictive. Chocolate also contains *phenylethylamine* which is a stimulant for people who are depressed mainly as a result of unbalances in their body chemistry. This is one reason for not being able to put down a box of chocolates.

Processing

Plain chocolate is made from cocoa beans that must be carefully chosen and roasted to give the characteristic bitter flavour

(a) The chocolate mix is made from chocolate liquor, to which is added sugar and cocoa butter

(b) A paste is produced by a special scraper-bladed mixer with heavy rollers. These rollers grind the paste into fine particles and are usually called *refiners*

(c) The final process, **conching**, is carried out in a large vat with a large roller that moves backwards and forwards. This is a slow

153

process, sometimes taking up to 48 hours at a temperature of around 65 °C. During this process the chocolate develops its full flavour and texture

(d) After conching, **tempering** is carried out to give a gloss to the product. This is achieved by cooling the chocolate until it begins to thicken and then re-warming it

(e) The liquid chocolate is then poured into moulds which set to produce the familiar bars

Milk chocolate must contain milk as well as the ingredients of plain chocolate. Usually sweetened condensed milk is used, to which is added chocolate liquor. This is then mixed and the resultant powder mixed with cocoa butter and processed to make chocolate as described above.

Nutrition

- Chocolate is primarily a source of *energy*, but contains some *proteins*
- The original cocoa beans contain significant amounts of *vitamins* particularly B complex and E. However, these are mostly lost in the finished product

Chocolate coatings

A large industry exists producing coatings or chocolate substitutes. The use of the word 'chocolate' in this case is debatable, as true chocolate must contain cocoa material – in most countries 'chocolate' must be made from pure cocoa nib.

Coatings are made for two purposes:

1 They are cheaper, as expensive cocoa butter is partly replaced by a vegetable fat

2 Pure chocolate is functionally unsuitable for some applications. For example, in ice-cream normal cocoa butter sets too hard, and the chocolate is brittle and flakes off. A softer vegetable fat is more suitable for this purpose

Cake coatings only require a short shelf-life. They are made from low fat cocoa (9–10 per cent cocoa butter), hardened palm kernel oil, sugar and milk powder (for lighter colour products). Lecithin (emulsifier) and flavour may be added. The coatings are made as for chocolate but the tempering stage is missed as the cake is covered with the melted coating at around 45 °C. The coating sets to a smooth finish with a good gloss. Fat bloom (white surface) readily appears on the coating towards the end of its shelf-life.

Sugar

Composition

Sugar is **sucrose** and pure table sugar is 99.9 per cent sucrose. For many years most sugar came from *sugar cane* but now the majority in the UK comes from *sugar beet*. After processing the sugars from both sources are identical.

Brown sugar is essentially a mixture of white sugar and molasses. *Demerara* sugar is raw cane sugar which has only the coarse impurities removed. *Icing sugars* are milled, crystalline, white sugars that are required to meet certain particle sizes.

Processing

Extraction from beet

(a) The beets, being swollen roots, are covered in soil and so efficient cleaning is necessary
(b) They are then shredded and the sugar extracted with hot water
(c) The sugar solution is separated from the pulp and refined as described below

Extraction from cane

(a) The cane is processed as soon as possible after cutting when it is crushed and mixed with water
(b) The sugar solution is separated from the pulp and purified using lime to precipitate impurities
(c) The sugar solution is then heated to drive off the water
(d) Crystals of sugar are produced within a syrup, the latter being known as 'molasses'
(e) A centrifuge is used to remove the molasses, which have several uses including the traditional one of rum-making
(f) The sugar crystals are then refined in a number of stages

Refining

(a) The sugar crystals are washed to remove further impurities, and the resulting solution subject to treatment with lime and carbon dioxide. The carbon dioxide reacts with the lime (calcium hydroxide) to produce calcium carbonate which slowly precipitates, removing with it impurities in the sugar solution

(b) The sugar solution is still brown and this colour is removed by passing the solution through a bed of bone charcoal

(c) The sugar solution is then evaporated under reduced pressure

(d) At a certain specific gravity the sugar solution is seeded with sugar crystals of the desired size (smaller ones for castor sugar) and the sugar begins to crystallize

(e) Surplus syrup is removed from the crystals by a centrifuge, and the crystals are sprayed with water and then dried by warm air

Nutrition

- Sugar is a supplier of *calories* and nothing else. A high intake can cause obesity and often tooth decay

- Work in the past has implicated sugar as a cause of heart disease and, more recently, as causing an increase in blood fat levels and hence heart problems. These theories have now been discarded and sugar has been vindicated on these issues

- Brown sugar is little different from white sugar and is not a 'healthier' product to eat. It is still refined and, although molasses contain *vitamins B*, *iron* and other *minerals*, little of these appear in brown sugar

Sugar confectionery

An enormous range of confectionery products is produced by controlling the state of sugar crystallization and the ratio of sugar to water in the product. Sugar crystals may be large or small, or the sugar may be non-crystalline and glass-like. Increasing the moisture and air content produces softer products.

- Seaside 'rock' is crystalline sugar in the form of one large crystal

- Fondants, for example chocolate egg fillings, contain very small sugar crystals
- If non-crystalline sugar is used the moisture content will vary from about 8–15 per cent. The latter level of moisture is found in soft products
- Very soft products, such as marshmallows, have high moisture contents and a considerable amount of air whipped into them

In many products **invert sugar** is added or produced by acid or enzymes (invertase). This gives extra sweetness but also, because of its high solubility, prevents recrystallization of the sucrose.

Instant desserts

Composition

Instant desserts must gel within a short time when added to milk or water. Various systems of gelling have been employed, including the use of *gelatine*, *pectin* and *pre-gelatinized starches*. The most popular types of instant desserts employ two systems which gel when added to milk:

(a) *Phosphates* and extra *calcium* gel the caseins in milk, often with the help of additional caseinate to give a firmer gel
(b) The gel is reinforced by pre-gelatinized starches which thicken immediately on adding milk or water

Instant desserts generally contain fat which is bound in a powder or encapsulated. This fat powder gives a better mouth-feel, but does not prevent the formation of a foam in the product as it is whipped during preparation. Special protein-derived agents are used to produce a foam by incorporating air during preparation. Whey proteins, for example, are sometimes used for their foaming properties. Emulsifiers are also employed to ensure good mixing of the fat powder.

Processing

- Often instant desserts are simple blends of dried ingredients. If the ingredient particles are of a similar size, blending is easy. Varying sized particles, however, are much more difficult to mix

- Fat powders are produced from emulsified fats in water, which are mixed with a carrier substance such as starch or gelatine, and then spray dried
- Well produced flavours are necessary as any with a 'synthetic' note will be noticeably inferior

Nutrition

- Instant desserts are not for those who dislike additives. The number of '*E*' numbers on a label can be quite noticeable
- The main contribution to the diet comes from the milk added, so these desserts are really a means of consuming milk, especially for children
- Catering products incorporate milk powder, so only water is added

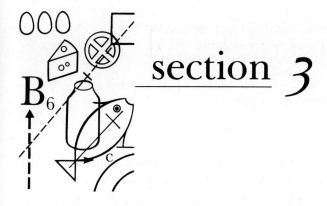

section 3

Glossary

This glossary is composed of words and terms commonly used in Food Science and Technology or related subjects. The definitions are not exhaustive but are intended to give a working understanding of each term.

Note: (i) All words appearing in bold type in Sections I and II are defined in this glossary;

(ii) Terms used in the glossary that have their own entry are set in italic.

Acids

Most foods are acidic. In fruits and vegetables citric and *malic acids* are the most common. In fermented products, such as yoghurt, *lactic acid* predominates. Acid is added to manufactured foods to improve flavour and to balance excessive sweetness.

Acraldehyde see *Acrolein*

Acrolein

Also known as *acraldehyde*. This is a breakdown product when *glycerol*, fats or oils are heated to a high temperature. It appears as a vapour with an acrid odour.

Additives

Chemicals, both synthetic and natural, which are used to give various functional properties to foods. In the quantities used, additives are edible but are not foods in their own right. Permitted additives within the EC are usually given *E* numbers.

Adenosine

A combination of adenine (a base) with the sugar ribose. Adenosine monophosphate contains one phosphate residue, diphosphate two, and triphosphate three. The latter, adenosine triphosphate (ATP) is important in the liberation of energy from foodstuffs. It is also involved in supplying energy to muscles during contraction.

Ageing

- of eggs
  Thick white of the egg becomes thinner, and membranes around the yolk weaken. Foaming properties of white improve with ageing.
- of flour
  Flour improves by keeping for several months. *Pigments* are bleached by oxidation, and *protein* quality is improved. Accelerated ageing is obtained by using oxidizing agents.
- of meat
  Enzymes break down large *protein* molecules after *rigor mortis* to make meat more tender. Free *amino acids* and *fatty acids* are produced to contribute to meat flavour.
- of wine
  A slow oxidation process that reduces astringency due to *tannin* in red wines. The production of *esters* is encouraged and a smoother wine with a bouquet is produced.

161

Alginates
Produced from giant kelp; sodium alginate is most common form. Widely used as additives, properties include: emulsifying, stabilizing, gel-forming, film forming and thickening.

Amino acids
Proteins are built up of amino acids which must contain an amino group (—NH$_2$) and a carboxyl group (—COOH). Amino acids are linked by *peptide* bridges (—CONH—) to form *peptides, polypeptides* and *proteins*.

- essential amino acids
  These must be included in the diet and cannot be synthesized in the body. Eight come in this category: valine, leucine, isoleucine, phenylalanine, threonine, methionine, tryptophan, lysine.

Children also require histidine and some people only sythesis arginine slowly.

Amylases
Enzymes which break down *starch*. α-amylase (or dextrinogenic amylase) attacks starch molecules randomly to produce *dextrins*. β-amylase systematically removes *maltose* from starch molecules. The two enzymes together are called *diastase*.

Amylopectin
One structural form of *starch*, the other being *amylose*. Composed of interconnected short chains of α-glucose to give a much branched, tree-like structure. Amylopectin is more difficult to *gelatinize* than amylose but forms stable gels resistant to *retrogradation*.

Amylose
A straight chain form of *starch* made of α-glucose units. The chain takes the form of a spiral or helix, with six *glucose* units per turn. Amylose readily undergoes *gelatinization* but after a time *retrogradation* occurs with the gel contracting and releasing water (*syneresis*).

Anaerobic respiration
Metabolism that occurs in the absence of oxygen. Compared with aerobic respiration it is much less efficient in producing energy and leads to the accumulation of products, such as alcohol, which may eventually have a toxic effect.

Anthocyanidins

Benzopyran derivatives with two ring structures (benzene derivatives) joined by a bridge. The basis of *anthocyanins* when combined with sugars. Substitution of different groups into the ring structures can lead to different coloured *pigments* in fruit, vegetables and flowers. An increase in —OH groups gives blue colours and increase in —OCH_3, gives red.

Their names indicate the colours they produce by association with flowers: eg

peonidin   – red
delphinidin – blue
petunidin   – mauve

Anthocyanins

Made from an *anthocyanidin* joined to a sugar or sugars such as *glucose*. Form a wide range of water-soluble *pigments* with colours ranging from blue to red. They are pH sensitive, like litmus paper, becoming red at lower pH and blue at higher pH. (See *anthocyanidins*.)

Anti-caking agents

Absorb moisture from dried foods without themselves becoming wet. Addition to powders ensures free-flowing characteristics eg salt. Common examples: magnesium oxide, silicates, calcium phosphates, salts of some long chain *fatty acids* such as stearic and palmitic.

Antioxidants

Prevent *rancidity* developing in fats by either absorbing oxygen or preventing chemical changes involved in rancidity. Antioxidants which absorb oxygen, include: *ascorbic acid* (vitamin C), gallates, and *tocopherols* (vitamin E). Chemical reactions involved in rancidity are prevented by BHA (butylated hydroxyanisole) or BHT (butylated hydroxytoluene). *Hydrolytic rancidity* cannot be prevented by antioxidants.

Ascorbic acid

Ascorbic acid is vitamin C in either its oxidized or reduced form. The latter is a powerful reducing agent and can be used as an *antioxidant*. Used in a number of products to protect flavours and fats against oxidation. Used in the Chorleywood bread process to speed dough development.

Naturally occurring in many fruits and vegetables, deficiency can lead to a range of diseases including bleeding of gums, bruising, internal bleeding and scurvy. Best sources are tropical fruits. Blackcurrants 200 mg/100 g, rose-hips 175 mg/100 g, oranges 50 mg/100 g. Potatoes supply a significant amount in the British diet because of the large quantity eaten, but contain only 8–30 mg/100 g, or less, when stored for several months.

Aseptic packaging
A previously sterilized product is filled into a sterile container. The container is sealed in a sterile environment to obtain a hermetic seal. Used for custards, ice-cream mixes and soups.

Aspartame
A new sweetner made from the combination of two *amino acids*, phenylalanine and aspartic acid. Nearly 200 times as sweet as *sucrose* with none of the after-taste of *saccharin*. Under acidic conditions and prolonged storage may break down.

Aspiration
A development of the ancient process of winnowing. A strong blast of air is used to separate food materials from contaminants, eg chaff from grain in wheat milling.

Avidin
An egg protein found in the white. Its anti-microbial properties as it binds the vitamin *biotin* make the latter unavailable to bacteria.

Bacteriophages
Viruses which attack bacteria only. They cause problems in *starter cultures* used in cheese-making by killing the bacteria in the cultures, thereby preventing development of acidity in the cheese. Abbreviated to 'phage.

Baking powders
Used to produce carbon dioxide in cake mixes to help give the required texture to the cake. Formed from sodium hydrogen carbonate (bicarbonate), a slow acting acid and a starch filler. Acids used include *cream of tartar*, calcium phosphate, disodium pyrophosphate and glucono-delta-lactone.

Bar coding
A code in the form of bars on a pack that can be read rapidly by electronic scanners at checkouts. The scanner is linked to a computer which feeds data back to the checkout and which prints the data automatically onto the till ticket. The code bars are split

into: prefix for the country of origin; manufacturer's number; item number and check digit.

Benzoic acid
A permitted *preservative* in a number of products. Like the benzoates (also permitted preservatives) it is used widely in soft drinks and similar products, but use is declining owing to allergic reactions in some people, and the public's reaction to additives in general.

Benzoic sulphimide see *Saccharin*

Benzopyran derivatives
Formed from two benzene rings and a joining bridge. Basis for coloured *pigments* such as *anthocyanins*, anthoxanthins and flavone derivatives.

Beri-beri
A disease caused by the deficiency of vitamin B_1 (thiamine). Symptoms include muscular weakness, palpitations, fever and heart failure. Outbreaks in the Far East have been caused by polishing rice, which removes main source of vitamin B_1.

Biosynthesised proteins
A *protein* produced by growing micro-organisms on suitable substrates, usually by a continuous *fermentation* method. Proteins have a good *amino acid* make-up but consumer resistance has been encountered.

Biotin
A *vitamin* synthesized in the human gut by bacteria. Deficiency is very rare. Egg protein *avidin* combines with biotin making it unavailable to micro-organisms.

Bitterness
A property of a range of organic and inorganic substances, particularly those containing magnesium, ammonium and calcium derivatives or quinine and *caffeine* in drinks.

Blanching
A short heat treatment carried out on vegetables prior to *canning*, *freezing* or *drying*. Main objectives are to inactivate *enzymes* and to shrink the product.

Efficiency of blanching tested by *peroxidase test*. Main methods are water and steam blanching. New developments include use of *microwaves*.

Blast freezing
Cold air (at around $-25\,°C$) is blown on to a food product of any shape to freeze it. The rate of freezing depends on the air temperature, its velocity used, and the surface area of the food exposed. Fluidized bed freezers are a development in which air is blown upwards to 'fluidize' the product as well as freeze it.

'Boil-in-the-bag' products
Food products prepared and packed into bags which can be immersed into boiling water for rapid cooking. High density polythene is commonly used, but new, thinner bags are available in linear low-density polythene.

Botulism
A dangerous disease with up to 70 per cent mortality rates, caused by the ingestion of a toxin produced by the organism *Clostridium botulinum*. The organism produces heat resistant spores which can survive in canned foods that are underprocessed. The spores require a temperature of $120\,°C$ and a pH less than 4.5 for complete destruction. The toxin, of which only minute quantities can be fatal, is destroyed above $63\,°C$.

Browning reactions
There are two main types:
- non-enzymic
    (a) *Caramelization* – breakdown of sugars at high temperatures to produce brown products
    (b) *Ascorbic acid* oxidation – slow process in some fruit juices leading to brown *pigments*
    (c) *Maillard reaction* – *reducing sugars* and *amino acids* or *proteins* react together eventually to form brown *melanoidin pigments*. The reaction is quicker at higher temperatures and high pH
- Enzymic
    In some fruit and vegetables the *enzymes, polyphenolases*, react with phenolic substrates in the presence of oxygen to produce brown *pigment*. Ideally substrates are diphenols, eg *catechol* in apples. Browning results from bruising, cutting, or some processing.

Browning can be prevented by using sulphites or sulphur dioxide, by inactivating *enzymes* or by removing one or more of the reactants eg by removing sugar to prevent the *Maillard reaction*.

Brucella
A genus of bacteria, of which *B. abortus* causes abortion in cattle

and can contaminate milk. Disease in humans is an undulating 'flu-like disease. Cattle now tested for brucella, and the organisms are usually destroyed in pasteurization.

Buffers
Control and stabilize the pH of food. Some food components act as buffers, eg *amino acids* and *proteins*. Weak *acids* and their salts are used eg *lactic*, citric, *malic* and *tartaric* acids.

Caffeine
A stimulant found in tea and coffee, considered to be addictive. Caffeine has been implicated as a possible cause of a number of disorders including nervous problems, but little has been proved. Coffee beans contain about 1 per cent caffeine. Referred to as an alkaloid drug; also called *trimethylxanthine*.

Calciferols
There are two slightly different calciferols: cholecalciferol (vitamin D_3) and ergocalciferol (vitamin D_2), and both are related to *cholesterol*. Produced by action of sunlight on the skin or found in fish oils.

About 2.5 μg per day is adequate; excessive can cause poisoning by damaging the kidneys. Involved with calcium absorption and deficiency leads to the disease *rickets*.

Calcium
Combines with phosphate to make up the structure of bones and teeth. Only about 40 per cent of calcium in food can be absorbed by the body. Main sources are dairy products, bread and some vegetables. Deficiency a problem in old age, particularly in women, causing the disease osteoporosis.

Canning
The process by which food is hermetically sealed in a container and then heat processed. The heat process is sufficient to kill all micro-organisms and their spores. A few spores may survive so the product is termed 'commercially sterile'; absolute *sterility* is very difficult to attain.

Caramel
Produced when sugars are heated above their melting point. Caramel is a range of brown substances, of variable composition. Some components may have a bitter flavour. It is used as a brown colour for foods and as a flavouring agent.

Caramelization

Process by which caramel is produced. A *browning reaction* in the absence of amino acid or proteins (see *Maillard reaction*).

Carbohydrates

Hydrates of carbon, made up of carbon, hydrogen and oxygen. Included in the groups are sugars, oligosaccharides and *polysaccharides*. The process of photosynthesis is responsible for producing carbohydrates in plants.

Carbonated drinks

Soft drinks to which is added carbon dioxide to give sparkle, better mouthfeel and some acidity. Carbonated drinks are sweetened, flavoured, acidified, coloured and often chemically preserved.

Carotenoids

A group of fat-soluble *pigments* ranging from yellow to red in colour. All are similar to carotene, which, as the name suggests, is found in carrots.

True carotenoids contain 40 carbon atoms, and are built up from 8 *isoprene* (C_5H_8) units. They are unsaturated hydrocarbon derivatives with conjugative double bond systems. Carotenoids with —OH groups present are called *xanthophylls*. Some have *pro-vitamin A* activity, eg β-carotene.

Case-hardening

A problem occurring in the dehydration of some foods, eg fruits, when a skin is produced on the surface which inhibits complete dehydration and retards rehydration. Usually caused by high temperatures, soluble solids and the denaturing of proteins on the surface of the produce.

Casein

The main milk *protein*, representing about 3 per cent of milk. Precipitated by *acid* at about pH 4.6 and coagulated by *rennet* in cheese making. A mixture of phosphoproteins with different properties. $α_s$-casein is precipitated by calcium, whereas κ-casein is insensitive to calcium.

Catechol

A substrate (a diphenol) for *polyphenolases* in apples which causes *enzymic browning*.

Cellulose
The most abundant *carbohydrate*, undigestible but a good source of fibre. Two types: crystalline and amorphous; the latter absorbs a considerable quantity of water and has been used in slimming foods. It is a *polysaccharide* composed of long chains of β-glucose, providing the main support for cells in plants.

Chilling
A method of extending the storage life of a product by lowering its temperature to between -1 and 8 °C, but not by freezing the product. Current research indicates that a temperature of below 4 °C is preferable to ensure that the product is safe from listeria infection.

'Chinese Restaurant Syndrome'
A feeling of dizziness and sickness apparently brought on by excessive consumption of *monosodium glutamate* (MSG), which was found to be used excessively in Chinese restaurants. However, MSG is naturally present in most foods, as it is a derivitive of the *amino acid* glutamic acid.

Chlorogenic acids
Complex organic acids found in fruit and vegetables. As they contain *phenolic* groups (—OH) they are ready substrates for *polyphenolase* enzyme activity to produce brown *pigments*. Similarly, they are found in *tannins* and coffee.

Chlorophyll
The *green* pigment of plants, involved in photosynthesis, the process by which plants manufacture carbohydrate. Under acid conditions the pigment is broken down to a greyish/brown colour, called *pheophytin*. It is stable under alkaline conditions, hence the addition of bicarbonate of soda (sodium hydrogencarbonate) to green vegetables when cooking.

Cholesterol
A widely distributed sterol found in animal tissue. The body can produce up to twice its normal requirements. It has been implicated in circulatory diseases, particularly as it can be deposited on the walls of blood vessels. Related chemically to vitamin D.

Clarification
Process of removing suspended or collidal material from liquid products, such as wine. Centrifugation, filtration and *enzymes* can be employed.

Climacteric
The peak of the rise in respiration shown in some fruit after harvesting. Climacteric fruit include: banana, avacado, mango and tomato. The rise in respiration is accompanied by changes associated with *ripening*.

Cold chain
A loose title for the course of events of transport and storage of frozen foods, from the *freezing* of foods, through storage and distribution, finally to the home.

Commercial sterility see *Sterilization*

Conching
A process used in traditional chocolate manufacture to help develop the right texture.

Controlled atmosphere (CA) storage
The storage of food products, particularly fruit and vegetables, in an atmosphere modified in some way, usually by changes in temperature and gases, particularly oxygen and carbon dioxide. Lower temperature and lower oxygen or high carbon dioxide levels are common.

Cryogenic freezing
The use of liquified gases for rapid *freezing* of small food items, eg raspberries and prawns. The liquified gas, usually nitrogen or carbon dioxide, is sprayed on to the food product. Rapid freezing is possible for small items only.

Curd proteins
Clotted *proteins* produced by the action of *rennet* on milk. *Casein* is modified by the rennet allowing *calcium* to react with the κ-casein, causing it to coagulate. The *whey proteins* are drained away to leave the curd which is transformed into cheese.

Curing
A term usually applied to meat which involves the development of colour, flavour and enhanced keeping qualities. Curing brine is added to meat and comprises salt, sodium nitrate, some sodium nitrite and sugar. Nitrate is converted to nitrite which combines with *pigments* in the meat to produce the characteristic pink/red colour.

Cutin
A general name for waxy substances found on the surface of fruit, eg on apples.

Cyanocobalamin
Vitamin B_{12}, essential for the formation of red blood cells. Deficiency disease (the inability to absorb the vitamin) is pernicious anaemia. Found in animal products, a particularly rich source being liver.

Cyclamate
An artificial sweetener, once popular but now banned. It was implicated as a possible cause of bladder cancer, but research carried out has subsequently been questioned.

D value
A *canning* term: the 'decimal reduction' value or 'D' value, refers to the reduction of bacterial spores to one-tenth of the original number in a canned product as a result of heat processing. For example, a D value of 4 means that it takes 4 minutes at 121 °C (250 °F) to reduce the number of bacterial spores to one tenth of the original number.

Dark cutting beef
A defect of beef caused by too high a pH owing to insufficient *lactic acid* being produced after slaughter. Oxygen is unable to penetrate the meat to produce the bright red oxymyoglobin; a dark purplish/red product is therefore produced. On cooking, the meat is dark in colour and often tough.

Demersal fish
Fish which are bottom feeders and are usually 'white', eg cod, saithe, sole, haddock and whiting.

Denaturation
Generally refers to the uncoiling of *protein* chains caused by heat, changes in pH, agitation and sometimes light. Slight denaturation can be reversible, and, in the case of *enzymes*, activity is regenerated.

Dextrins
Breakdown products of *starch* which are soluble, long chains of α-*glucose* units. They are formed when bread is toasted and are often used as edible adhesives.

Dextrose

Alternative name for α-D-glucose. Name originates from the ability of dextrose solution to rotate a plane of polarised light in a clockwise or dextro-rotatory fashion. Given the symbol (+), as opposed to laevorotatary, as in *fructose*, which is given the symbol (−).

Dextrose equivalent value (DE)

A value denoting the degree of conversion of *starch* into *glucose*. The higher the DE, the more glucose (*dextrose*) present.

Diastase

The combination of α- and β-*amylases* which break down *starch*.

Dielectric heating

A form of heating produced in a material when it is subjected to an alternating electric field. Heating is caused by molecular friction due to rapid movement of molecules, such as water, in the alternating electric field. *Microwave* heating is similar, but is caused by electromagnetic radiation.

Diglycerides

Emulsifying agents, composed of *glycerol* and two *fatty acids*. Not as effective as *monoglycerides*.

Dipeptides

Two *amino acids* linked together by the *peptide* link or bridge (—CO—NH—). Formed in cheese *ripening* in some varieties.

Disaccharides

Sugars, such as *sucrose*, *lactose* and *maltose*, formed by the combination of two *monosaccharide* units with the elimination of a molecule of water.

$$sucrose \rightarrow fructose + glucose$$
$$lactose \rightarrow galactose + glucose$$
$$maltose \rightarrow glucose + glucose$$

Disulphide bridge

A bridge or link formed between *protein* chains from the *amino acid* cysteine. The bridge (—S—S—) gives elasticity to dough during bread-making.

Diterpenes

Derivatives of isoprene (C_5H_8), all diterpenes are based on four isoprene units ($C_{20}H_{32}$). (see *Terpenoids*)

Drip
The liquid which exudates from a frozen product on thawing. Found in fish and meat if frozen slowly.

Drying (dehydration)
Removal of water from a food to preserve it.

- freeze drying
  Food frozen then subjected to strong vacuum to sublime the ice and leave food dry; expensive.
- roller drying
  Used for liquids; product in contact with hot roller or drum; relatively slow drying for milk or whey.
- spray drying
  Liquid sprayed into chamber and met by a blast of hot air to produce a fine powder, eg coffee.
- fluidized bed drying
  Hot air blown upwards into a bed of product, which then acts as a fluid, and dries fairly rapidly.

Electrophoresis
The movement of electrically-charged particles when an electric current is passed through a solution. Used to separate *proteins*. Fish can be recognized by the bands of proteins produced by electrophoresis of fish flesh within special gels, each fish species having its own unique bond formation.

Emulsifying agents (emulsifiers)
Substances which enable the production of a stable dispersion of oil in water or vice versa. Examples include: *glyceryl monostearate* (GMS), *lecithin*, egg yolk, and *whey protein*. Some are more soluble in oil than water, but all orientate themselves at the interface between oil and water and prevent droplets of oil from coalescing and separating.

Enzymes
These substances are naturally occurring catalysts found in the cells of plants or animals. They exhibit the properties of *proteins* and may be very specific in their action. If not inactivated they can produce undesirable changes in processed foods, such as changes in flavour, colour or texture.

Enzymic browning see *Browning*

Erucic acid
A *fatty acid* (22 carbon atoms) found in rape seed oil. Implicated as
a cause of heart disease. Modern varieties are bred to be free of
the acid.

Essential amino acid see *Amino acid*

Essential oils
Natural oils which contribute to the flavour of foods, particularly
fruits. Name is derived from 'essence' – they are not 'essential' for
life. They are the basis of naturally occurring food flavours, such
as citrus oils.

Esters
Formed from the reaction between organic acids and alcohols.
Many have fruit-like flavours, eg ethyl ethanoate (acetate) which is
like pear-drops.

Ethylene (ethene)
A gas which acts as a plant hormone, stimulating the *ripening*
process and the increase in respiration rate in *climacteric* fruit.
Thought to be produced by the breakdown of the *amino acid*,
methionine. The gas can be added externally to ripen fruit, eg
bananas.

Extraction rate of flour
Indicates the yield of flour obtained when milling wheat.
Wholemeal flour is 100 per cent extraction ie the whole wheat
grain. White flour is about 72 per cent extraction as the bran and
germ have been removed.

Extrusion
A modern process of making snack foods, particularly corn and
potato based products. The food material is heated under pressure in a barrel-shaped piece of equipment with either a single or
double internal revolving screw. The revolving screw takes the
product to a dye from which the product is released. The sudden
pressure drop, on release from the dye, makes the product
expand rapidly to produce a light porous texture.
    Various dyes are used to give products of different shapes and
sizes.

F value
A term used in canning to denote the overall lethal effect of the
heating process. It refers to the number of minutes at which the
product is kept at 121 °C (250 °F) that give an equivalent steriliz-

ing effect to carrying out the whole sterilization process. The exact time is determined by that required to achieve 'commercial sterility', when the bacterial population has been reduced by 12D, ie down to 10^{-12} of the original level. (See *D-value*.)

Fatty acids

Organic acids containing a carboxyl group (—COOH); long chain acids found in fats combined with *glycerol*.

- Essential fatty acids
  Fatty acids required by the body, eg linoleic (the main one), linolenic and the third, arachidonic, which is produced by the body.

Fehling's Test

A test for reducing substances, particularly *reducing sugars* such as *glucose, fructose, maltose* and *lactose*. Two solutions are mixed then boiled with the sample. Fehling's solution 1 is copper sulphate and solution 2 is a mixture of sodium hydroxide and tartrate *buffer*. When mixed, reducing substances turn the blue solution to a brick red colour due to the production of copper 1 oxide.

Fermentation

The metabolism of organic compounds in the absence of oxygen. Yeast ferments sugar to produce alcohol. *Lactic acid* bacteria ferment *lactose* in milk in yoghurt manufacture. Fermentation produces energy for the organism but inefficiently compared with aerobic respiration.

Fibre

The indigestible parts of food, generally consisting of *cellulose, hemicellulose* and *pectin*. Important in maintaining proper functioning of the digestive system. The NACNE report recommends a minimum of 30 g per day.

Filled milk

A dried milk containing vegetable fat instead of butter-fat. Has better keeping qualities than whole-fat milk powder. Used by the catering industry.

Filter aid

A substance made up of large, inert particles that is mixed with the liquid to be filtered to prevent holes in the filter becoming blocked. The filter aid forms a porous structure on the filter allowing the liquid to flow through freely. Examples are kieselguhr, paper pulp and carbon.

Flavonoids
A large group of compounds found to be a basis of *pigments*, such as *anthocyanins*, and also some bitter components of foods.

Flavours
Flavours are detected by the nose and therefore volatile substances contribute most noticeably to flavours, eg fruit flavours. *Essential oils* from fruits are main flavour components. Synthetic flavours often contain *esters*. Nature-identical flavours contain the same chemical substances that are found in the natural product or flavour.

Flavour enhancers
Substances which enhance or improve a flavour. *Monosodium glutamate* (MSG) has been widely used in many products, particularly meat, fish and vegetable products. *Ribonucleotides* are similar to MSG but sometimes more effective. Only very small amounts are needed. Their action may be through the stimulation of the taste buds.

Flavour profile
A taste-panel procedure whereby a flavour is broken down into a number of factors or characteristics. The characteristics are often rated on a scale from 1 to 10.

Fluidized bed drying see *Drying*

Fluidized bed freezing see *Blast freezing*

Folic acid
A vitamin essential in the synthesis of certain *amino acids* and similar components. Deficiency causes a form of anaemia. Widely found in foods, eg liver, vegetables and yeast.

Free radicals
Highly reactive chemical species formed from molecules with unpaired electrons. They are involved in initiating and perpetuating the reactions which cause fat to become *rancid*. The *antioxidants* BHA and BHT produce stable free radicals and stop the *rancidity* reactions.

Freeze drying see *Drying*

Freezer burn
The drying out of the surface of a frozen food, particularly meat or fish, which is unwrapped. The problem is caused by the

sublimation of ice crystals from the surface directly to water vapour. The thawed product is tougher or drier to the palate.

Freezing
The conversion of water from the liquid state to solid ice. The thermal arrest period of freezing is the time taken for all the available water to be formed into ice. Common methods:

Immersion – food placed in a very cold brine
Plate – freezing by contact with cooled flat surface
Blast – cold air blown over the product to freeze it
Cryogenic – liquid gases, nitrogen or carbon dioxide are sprayed over the food.

Freshwater fish
Fish that live in non-salt water and are unable usually to live in sea water, eg trout, pike, perch.

Fructose
Also known as *laevulose*. Rotates a plane of polarized light anti-clockwise, hence 'laevorotatory' (−). A very sweet sugar (*monosaccharide*) about 1.6 times as sweet as *sucrose*. Found in *invert sugar*, honey, jam and confectionery.

Fuller's earth
Used in the refining of oils to remove *pigments* such as *carotenoids*. Porous colloidal aluminium silicate is the main constituent.

Galactose
A *monosaccharide* and *reducing sugar*, found linked to *glucose* in *lactose* (milk sugar). It has a six carbon ring structure similar to *glucose*, except for the position of the hydroxyl group on carbon atom four. Has about one-third the sweetness of *sucrose*.

Galacturonic acid
An acid derivative of *galactose* with a carboxyl group (—COOH) on carbon atom six. A component of *pectin*. Pectic acids are predominantly galacturonic acids in long chains.

● methyl galacturonate
Methyl *ester* formed from galacturonic acid. Found in *pectin*. *Pectinic acids* are predominately methyl galacturonate with some galacturonic acid in long chains.

Gas chromatography

A type of chromatography using columns packed with a stationary phase through which is passed a carrier gas. Volatile substances are separated by the procedure and detected by special detectors, usually flame ionisation detectors. Useful for separation of *flavour* components.

Gelatinization

The process by which a gel is formed. In the case of *starch* a large quantity of water is absorbed and the starch eventually cross-links to form a three-dimensional network. Starches gelatinize at certain temperatures. Starches higher in *amylose* gelatinize at lower temperatures than those rich in *amylopectin*. The latter produces more stable gels.

Glucose

A *monosaccharide* containing six carbon atoms, also called *dextrose*. A *reducing sugar* of about 70–80 per cent the sweetness level of sucrose.

Glucose syrup

A colourless syrup produced by hydrolysis of *starch*. Of variable composition including *glucose*, *dextrins* and *maltose*. Used as a sweetening agent in confectionery.

Gluten

The *protein* of wheat flour and, to a lesser extent, rye. Composed of two types of protein, gliadins and glutenins. Responsible for the extensibility and elasticity of dough in bread-making. Strong flours are richer in gluten. (See *wheat proteins*)

Glycerine see *Glycerol*

Glycerol

Commonly known as *glycerine*. A clear, viscous, sweet liquid. Chemically a trihydric alcohol, with three carbon atoms. Combines with *fatty acids* to form *triglycerides* which make up fats.

Glycerol monstearate (GMS)

An *emulsifier*, made from *glycerol* and one fatty acid, stearic acid. Commonly used in the food industry as a general purpose *emulsifier*.

Glycogen

A *polysaccharide*, similar to *amylopectin*, which is found in animal tissue. Composed of α-glucose units and is broken down in the muscles ultimately to yield energy for muscular activity. On the death of the animal it is converted to *lactic acid* which lowers the pH of the meat.

Glycosides
Substances made up of a molecule to which is attached a sugar molecule; eg anthocyanins contain sugars attached to an anthocyanidin molecule.

Glycosidic links
An oxygen bridge between two sugar molecules. *Maltose* is made of two *glucose* units joined by a glycosidic link between carbon atom one on one molecule, and four on the other. Hence the term '1–4 glycosidic link'.

Grading
Grading, or quality separation of foods, depends on a number of characteristics, eg size, shape, colour and freedom from blemish. Often confused with *sorting* which depends on one characteristic, eg size.

Gram stain
A staining technique used to distinguish groups of bacteria. Those which retain the stain (crystal violet) are termed gram positive, eg *Bacillus, Clostridium*. Those which lose the stain are termed gram negative eg *Salmonella, Pseudomonas*.

Gums
A wide range of *polysaccharides*, some of complex composition, which absorb large quantities of water and act as *stabilizers* and *emulsifiers*. Examples include, *alginates*; gums arabic, tragacanth, guar, and xanthan gum. Used in many food applications to prevent water separation and to thicken.

Haemoglobin
Red pigment of blood, composed of an iron containing haem structure joined to a *protein*, globin. Haemoglobin combines with oxygen from the lungs and transports it to the tissues. Iron deficiency (anaemia) results in the inability to produce enough haemoglobin.

Hemicellulose
A group of *polysaccharides* found in plant cell walls of variable composition. All are soluble in alkali and can be broken down to sugars, which are pentoses (contain five carbon atoms) eg *xyloses*, and some sugar derivatives eg *uronic acids*.

High-temperature short-time (HTST)
Heat processes such as *pasteurization* and *sterilization* of milk using plate heat exchanges. Process allows rapid heating of product with minimal flavour changes but full bacteriocidal effect of the heat.

High viscosity juice
Fruit juice that is thicker than normal juices (*low viscosity juices*) such as lemon juice, and has considerable suspended material thus rendering it opaque, eg tomato juice.

Homogenization
Process applied to milk to break down fat into small stable droplets which do not cream off. Normal method is to use a pressure homogenizer which forces the milk through a small orifice under pressure.

Humectants
The opposite of *anti-caking agents*, since they are used to keep products moist. They release moisture slowly to the product as it dries out. Examples include, *glycerol*, sorbitol, sodium and potassium lactate.

Hydrogen bonds
An electrostatic attraction between oxygen and hydrogen atoms. The oxygen has a slight negative charge and the hydrogen a positive one. Occurs only momentarily but between millions of atoms, so the overall effect is considerable. Involved in binding water to *proteins*, in the *gelatinization* of *starch* and in most food products.

Hydrogenation
The process by which unsaturated fats or oils are converted to saturated fats. The process involves passing hydrogen gas through the heated fat in the presence of a nickel catalyst. Used extensively in margarine and shortening manufacture.

Hydrolytic rancidity
Fats in the presence of water break down to release *fatty acids* from the *glycerol* in their constituent *triglycerides*. The process is accelerated by lipolytic enzymes (*lipases*) and micro-organisms, especially some moulds. Short-chain fatty acids, only, are of significance because of their odour.

Hydrostatic retorts
Large, continuous autoclaves used for *canning*. Operate with hot

water and steam. The head of water maintains this steam pressure. A moving belt takes the cans through the water pre-heating stage, then to steam heating and finally cooling. High rates of production can be achieved, but high capital outlay is involved.

Hypobaric storage
A method of storing food, particularly fruits and vegetables, that relies on reducing the atmospheric pressure in the store. In fruits it delays *ripening* by lowering the oxygen and ethene (*ethylene*) levels. It is expensive and difficult to maintain. Largely experimental and little used commercially.

Icing
A method of chilling, used particularly for fish. Melting ice is better as latent heat is absorbed, so a greater cooling effect is achieved. Tropical fish keep better by icing than temperate fish.

Immersion freezing
An old method of *freezing* using super-cooled brines. A slow method with little control, rarely used nowadays.

Improvers (for flour)
Chemicals that are used to accelerate the *ageing* of flour and improve its bread-making potential. The *protein* of the flour is affected and *disulphide bridges* produced. Improvers are oxidizing agents, eg ammonium persulphate and potassium bromate.

Individually-quick-frozen (IQF)
Rapid freezing, eg by *fluidized-bed freezing*, of individual foods such as peas. A continuous method giving a higher quality product.

Inositol hexaphosphoric acid see *Phytic acid*

Instantization
The alteration of a food to make it immediately dispersible and soluble in water. Milk powder is instantized by re-wetting, or not drying completely, so that powder particles stick together to form granules which readily dissolve in water.

Interesterification (of fats)
The process by which the positions of the *fatty acids* attached to *glycerol* in the *triglyceride* of fats are altered. The fat is heated in the presence of a catalyst, sodium ethoxide. Normally used to improve the creaming properties of fats to be blended for shortenings.

Inversion (of sucrose)
The splitting of *sucrose* into its component *monosaccharides, glucose* and *fructose*. Sucrose rotates a plane of polarized light clockwise, hence (+); *glucose* and *fructose* together rotate in an anticlockwise direction (−). The change in the sign from (+) to (−) is 'inversion'.

Invert sugar
A mixture of equal amounts of *glucose* and *fructose*. As fructose is strongly laevorotatory (−) and glucose only slightly dextrorotatory (+), invert sugar is laevorotatory (−) overall. Similarly, because of the high sweetness of fructose, invert sugar is sweeter than sucrose.

Occurs naturally in honey, jams and fruit juices. Produced in or added to confectionery for sweetness and also, because of its high solubility (absorbs water readily), to prevent water being used by sucrose to reform sugar crystals.

Iodine value
A measure of the unsaturation of a fat. A special preparation of iodine (Wij's solution) is used to determine the number of double bonds in fats. Higher iodine values indicate more unsaturation, eg butter IV around 29, lard 59, cottonseed oil 110.

β-Ionone ring
The ring structure at the ends of a *carotenoid* structure. An unbroken ring is necessary for a carotenoid to have *pro-vitamin A* activity.

Irradiation
The application of ionising radiations to kill bacteria in foods. Radiations are produced from decaying sources such as Cobalt 60 and Caesium 137, or by electron accelerators. There is no risk of residual radioactivity. Must *not* be confused with radioactive contamination or radioactivity.

Iso-electric point (IEP)
Refers to the pH at which a *protein* or *amino acid* has an overall net charge of zero; in other words, all negative charges are balanced by all positive charges. At the IEP the properties of the protein are at a minimum, eg solubility, electrical conductivity and viscosity. The IEP of *casein* is 4.6, hence its precipitation as milk sours.

Isoprenoid derivatives
Compounds built up of isoprene units (C_5H_8). *Monoterpenes*

182

found in *flavours*, such as *essential oils*, are made from two isoprene units. *Carotenoids* are made from eight isoprene units.

Lacquer (in cans)
A resinous coating on the inside of cans which is hardened by heat. The lacquers vary in composition according to function. The main types resist acid attack and others prevent the formation of sulphides from the reaction of sulphur with metal from the can.

Lactic acid
Produced by the *fermentation* of *lactose* in milk giving acidity and the flavour of sour milk. Produced by lactic acid bacteria, eg *Lactobacillus* in pickles. Important in meat, as *glycogen* is converted to lactic acid post mortem, and is responsible for a pH of meat of around 5.4.

Lactose
Only found in milk, at about 4.8 per cent. A *disaccharide* and reducing *sugar*, made from *galactose* and *glucose*. Only about 16 per cent of the sweetness of *sucrose*.

Laevulose see *Fructose*

Lecithins
Fatty substances belonging to the *phospholipids*. They are made of *glycerol, fatty acids* (at least one unsaturated), phosphoric acid and a nitrogenous base, choline.
  Naturally occurring emulsifying agents found in egg yolk, milk, soya and peanut. Many applications where natural *emulsifiers* are desirable, eg chocolate, mayonnaise. Commercially they are produced mainly from soya bean.

Lignin
A complex high molecular weight substance similar to *carbohydrates*. As plants age it is deposited on cell walls to toughen them. Found in old vegetables as a woody texture or as stringiness in runner beans.

Lipases
Enzymes (lipolytic) which break down fats. Usually they release *fatty acids* from their attachment to *glycerol*. Free fatty acids are measured as an indication of lipase activity.

Lipids
General term for fats, waxes and oils. Includes *phospholipids* such

as *lecithin*. Most are combinations of *glycerol* and *fatty acids*.

Listeria

A group (genus) of bacteria that can give rise to a number of diseases including meningitis. *L. monocytogenes* has caused concern as it has been found in a number of chilled food products such as soft cheese.

Low viscosity juice

Clear, thin fruit juice with little or usually no suspended matter, eg lemon juice, apple juice, grape juice (cf *high viscosity juices*).

Lycopene

The red *carotenoid pigment* of ripe tomatoes. It does not possess *pro-vitamin A* activity. Unlike most carotenoids, it is synthetized during *ripening* as chlorophyll is degraded.

Maillard reaction

A *browning reaction*, which is *non-enzymic*, resulting from the initial condensation of *reducing sugars* and *amino acids* or *proteins* (amino group). It occurs as the result of heating during cooking and processing, but also takes place slowly in some stored products, such as dried milk, if their moisture levels are above a certain level (around 5 per cent). The reaction is accelerated at higher pH and retarded by the addition of acid or sulphur dioxide.

In some foods it causes a reduction in nutritive value as essential amino acids can be bound up into complexes. Brown pigments formed are called *melanoidins*. Some immediate products and melanoidins contribute to flavour and have been used in producing synthetic meat *flavours*.

Malic acid

A common fruit acid, found in, for example, apples, plums and tomatoes.

Maltose

A *disaccharide* and *reducing sugar* composed of two *glucose* units. Produced by the malting process from *starch* in barley grains. Its sweetness is about 30 per cent that of *sucrose*.

Melanins see *Melanoidins*

Melanoidins

Also called *melanins*. A general name for the brown *pigments* produced during *browning reactions*, eg the *Maillard reaction*.

Metmyoglobin
The brown pigment of meat when it is old or cooked. The iron of *myoglobin* is changed by oxidation from Iron II to Iron III. It does not have the capacity to carry oxygen.

Micelles
Small bundles of *casein* molecules that occur in milk. The calcium sensitive (α_s casein) is protected by the calcium insensitive (κ-casein). On coagulation, eg in cheese-making, the micelles are modified and cross-linked.

Microwaves
Electromagnetic *radiations* having frequencies in the range of 3–300 000 MHz, although the most commonly used frequencies are between 915 and 2 450 MHz. The microwaves cause molecules to vibrate which generates heat. Only molecules that are irregular and have electrical dipoles (ie positive and negative charges) are affected – water is the most important molecule involved.

Microwave ovens are still the most common application. Commercial applications include thawing of frozen blocks and *blanching*.

Modified atmosphere
Refers to storage or packaging of products, where the atmosphere has been modified in some way, usually by decreasing oxygen and increasing carbon dioxide or nitrogen.

Now used particularly for modified atmosphere packaging where packs are gas flushed. Applications include fish products, fruit and vegetables, spoilage has been reduced and shelf-life increased.

Monoglycerides
Composed of *glycerol* and one *fatty acid*. The remaining part of the glycerol can dissolve in water and the fatty acid chain in fat, thus the monoglyceride is able to act an an emulsifying agent. A common example is *glyceryl monostearate*.

Monosaccharide
Group name for the simplest sugars ranging from three to seven carbon atoms. The most common contain six carbon atoms (hence 'hexose'), eg *glucose*, *fructose* and *galactose*.

Monosodium glutamate (MSG)
A *flavour enhancer*, found naturally in soy sauce, that occurs in

many foods. For many years MSG has been added to food products particularly meats and vegetables. It brings out meat flavour, and tends to round off and suppress undesirable flavours. Excess MSG intake has been shown to cause dizziness and sickness – known as the *'Chinese Restaurant Syndrome'* or 'Kwok's disease'.

Monoterpenes
Derivatives of isoprene (C_5H_8). All monoterpenes are based on two isoprene units ($C_{10}H_{16}$). Many are alcohols, aldehydes and ketones. They are constituents of *essential oils* and are to be found in many naturally occurring *flavours*. Monoterpenes are fairly unstable and can undergo changes in processing which lead to noticeable flavour changes. (see *Terpenoids*)

Myoglobin
Similar to *haemoglobin* in blood, but only one-quarter of the latter's molecular weight as it contains one haem and not four, as in haemoglobin. Responsible for storage of oxygen in muscles where it is converted to oxymoglobin, the bright red colour of meat.

Myoglobin itself is a duller red and can be oxidized to brown *metmyoglobin* when the Iron II atom in the haem part of the molecule is oxidized to Iron III.

Naphthoquinone
The main part of vitamin K. There are a number of *vitamins* related to this, called 'substituted naphthoquinones'. They are essential for the working of the blood-clotting system.

Niacin see *Nicotinic acid*

Nickel
A catalyst used in a finely divided state to facilitate the hydrogenation of unsaturated oils when converting them to harder fats. Used in margarine and fat manufacture.

Nicotinamide see *Nicotinic acid*

Nicotinic acid
Also known as *niacin* and can exist as *nicotinamide*. A *vitamin* of the B complex. Deficiency leads to *pellagra*, a mental disorder leading to insanity. Found in many foods including meat, cereal germ, yeast and liver. It is added to flour.

Nitrosamine
A carcinogenic substance implicated in cancer of the throat.

Formed from nitrite reacting with secondary amines. The former may come from bacon or cured meats. Chances of this occurring are slight.

Nitrosomyoglobin
The name given to the pink derivative of *myoglobin* produced by the action of nitrite during the curing of bacon and similar products.

Non-climacteric fruit
Fruit (and most vegetables) which do *not* show a rise in respiration rate after harvesting. Their *ripening* and respiration is gradual over a long period.

Non-enzymic browning see *Browning*

Novel proteins
Proteins produced by new methods, eg mycoprotein from fungi.

Nutritive additives
Vitamins and minerals added to certain foods such as flour, that are convenient carriers to ensure a balanced diet. Margarine, for example, has vitamins A and D added, whilst added to white flour are *calcium*, iron, and the vitamins *thiamine* and *nicotinic acid*.

Oils
Fats which are liquid at room temperature. Usually contain a greater number of unsaturated fatty acids which lower the melting point.

Oleoresins
The non-volatile flavour constituents of spices and herbs. They are normally extracted by solvents such as acetone (propan-2-one). For industrial use oleoresins are mixed with inert carriers, such as *starch*, salt or *dextrose*, for to enable easy mixing into foods.

Optical activity (in sugars)
Sugars and some *acids* have the ability to rotate a plane of polarized light, either clockwise (+) or anti-clockwise (−). Clockwise rotation is referred to as 'dextrorotatory', eg a solution of *glucose*; anticlockwise is 'laevorotatory', eg as in a solution of *fructose*. This optical activity can be observed using a polarimeter.

Oxalic acid
An organic *acid* containing two carboxyl (acid) groups. Found in rhubarb leaves, spinach and some chocolate. Toxic in high quantities; can bind iron and other minerals.

Oxidative rancidity
Occurs in unsaturated fats and *oils* and starts adjacent to the double bonds. The reaction is initiated by the presence of metals (particularly copper and iron), ultra-violet light, and high temperatures. Highly reactive *free radicals* are involved in the reactions.

Oxidation of the double bonds occurs to produce hydroperoxides which breakdown to produce the odour of rancid fat. Prevented by the use of *antioxidants*.

Pale soft exudate (PSE)
Refers to the fluid expelled from pork muscle, post mortem, due to a rapid fall in pH. Due mainly to hereditory factors in the species of pig. The pork meat as a result is soft and much paler in colour owing to loss of water-holding capacity and destruction of blood *pigments*.

Panthothenic acid
A *vitamin* needed for the metabolism of fats and *carbohydrates*. Found in many foods, so deficiency should never arise.

Pasteurization
A heat process intended to kill pathogenic micro-organisms and some spoilage organisms. Generally extends shelf-life but does not achieve any degree of sterility. Milk is pasteurized at 72 °C for 15 seconds.

Pectic
A complex *polysaccharide* found in plant cell walls and between cells in the middle lamella. Used with sugar and acid to form jam.
  Various types:

- Protopectin
  parent pectic substance; a large macro-molecule found in unripe plant products.

- Pectinic acids
  (normal *pectin*) made from long chains of *galacturonic acid* and methyl galacturonate – usually more than 50 per cent of the latter. Gels with sugar and acid in preserves.

- Pectic acids
  found in very ripe fruit. As *pectinic acids* but mainly free of methyl galacturonate. Will not gel with sugar and acid, but will gel with *calcium* and other metal ions.

- Rapid-set pectin
  pectinic acid with high methyl galacturonate content; sets

rapidly, so used to suspend fruit in jam.

- Slow-set pectin
  used for jams which are pumped or filled into cakes or pastries.
  Slow setting ensures operation carried out without gelling.

Pectic acid see *Pectin*

Pectinic acid see *Pectin*

Pelagic fish
Fish living near the surface of the sea. Oily fish with up to 20 per cent oil, eg herring and mackerel.

Pellagra
Deficiency disease caused by the absence of *nicotinic acid* in the diet. Can result in skin problems and mental disorders.

Penicillium
Group of common moulds which spoil many foods. *P. roquefortii* is used to give blue veins and flavour of blue cheeses.

Peptide bond
Link between *amino acids* in forming *dipeptides*, *polypeptides* and *proteins*. Formed from condensation of amino group (NH$_2$) and carboxyl group (—COOH), hence peptide bond (—CONH$_2$—).

Peroxidase test
A test for the efficiency of *blanching*. Peroxidase is a very heat resistant *enzyme* so its *denaturation* by heat can be assumed to have included the denaturation of other enzymes present in the food.

A sample of blanched product is tested by adding a drop of hydrogen peroxide followed by a drop of guaiacol. The peroxidase breaks down the hydrogen peroxide to release oxygen which turns the guaiacol a brown colour, usually within 30 seconds to 1 minute. No colour change is observed within this time if the peroxidase is denatured.

Usually performed every 30 minutes after blanching peas, for example, in a freezing or canning plant.

Phage see *Bacteriophage*

Phenolases see *Polyphenolases*

Phenol oxidases see *Polyphenolases*

Pheophytin
A grey-brown substance produced from *chlorophyll* in cooked or processed foods under acidic conditions. Hydrogen ions replace

the magnesium atom in the centre of the chlorophyll structure. The formation of pheophytin is prevented by making cooking water slightly alkaline, eg by adding sodium hydrogen carbonate (bicarbonate of soda).

Phosphate cross-bonded starch see *Starch*

Phosphatides see *Phospholipids*

Phospholipids
Fatty substances containing *glycerol, fatty acids*, phosphoric acid and a nitrogeneous base. A common example is *lecithin*, a natural *emulsifier*. Also called *phosphatides*.

Phytic acid Chemically: *inositol hexaphosphoric acid*. Occurs in the bran of cereals, peas and beans, and has the ability to combine with *calcium* and iron in other foods, hence the addition of calcium carbonate to flour.

Phytosterol
A general name given to plant sterols, similar to *cholesterol*, but not implicated in the latter's undesirable effects.

Pigments
Generally refers to naturally occurring colours. Three main groups:

Chlorophylls	– green
Carotenoids	– yellow to red
Benzopyran derivatives (mainly *anthocyanins*)	–red to blue

Plate freezing see *Freezing*

Polarimeter
An instrument to measure the degree of rotation of polarized light caused by a solution of a sugar, *amino acid* or organic acid.

Polypeptides
Long chains of *amino acids*; combine to form *proteins*.

Polyphenolases
Also known as *phenolases* and *phenol oxidases*. Naturally present *enzymes* in fruit and vegetables which, on damage to the cells, react with diphenols present, in the presence of oxygen, to produce brown *pigments*. This process is known as *enzymic browning*.

Substrates are usually diphenols such as *catechol* in apples. The enzymes react more slowly with monophenols such as tyrosine in potatoes.

Their action is prevented by heat, sulphur dioxide, acids and exclusion of oxygen.

Polysaccharides
Long chains of *monosaccharides* (sugars) joined together by *glycosidic links*. *Starch* is a chain of α-glucose units, *cellulose* is made from β-glucose units. Simple polysaccharides are made from one monosaccharide type, whereas in complex polysaccharides different units are involved.

Pre-gelatinized starches see *Starch*

Preservatives
Substances capable of preventing the spoilage of food by the action of micro-organisms. There is a permitted list for use in foods, the main two being *sulphur dioxide* and *benzoic acid*.

Preserves
A range of food products preserved by the osmotic effect of high sugar concentrations. Main examples are: jams, marmalades, conserves and candied fruit.

Proteins
Essential constituents of all cells, basically composed of carbon, hydrogen, oxygen and nitrogen. They are very large molecules built up of numerous combinations of *amino acids*. Proteins of high biological value contain all the essential amino acids.

Protopectin see *Pectin*

Pro-vitamin A
A *carotenoid* that can be converted in the body to vitamin A. Main examples include α-, β- and γ-carotenes. They must contain an intact β-*ionone ring* at the end(s) of their chain to qualify for pro-vitamin A activity.

Pyridoxine
Vitamin B_6, involved in *amino acid* metabolism in most animals. Widely present in foods so deficiency does not usually occur.

Radiations see *Irradiation*

Rancidity
A chemical change in fats and oils brought about by either oxidation or hydrolysis. Leads to the production of odours and flavours (caused by aldehydes and ketones) associated with the

deterioration of fat. In the case of *hydrolytic rancidity*, short chain-free *fatty acids* produce the odour.

See *Hydrolytic rancidity* and *Oxidative rancidity*

Rapid dough processes
The traditional dough process is accelerated by the use of mechanical mixing and oxidation. The dough structure is developed and fixed quickly. Little *fermentation* occurs relative to traditional processes. Main example: Chorleywood Bread Process.

Rapid-set pectin see *Pectin*

Reducing sugars
Sugars which contain a potential aldehyde group (—CHO). Some examples are: *glucose, maltose* and *lactose*. Some sugars, such as *fructose*, contain a ketone group but are still reducing sugars. Tested for by their ability to breakdown *Fehling's solution*.

Reducing sugars react with amino acids and proteins in the *Maillard browning reaction*.

Refrigerated sea water (RSW)
Used in some countries to chill freshly caught fish instead of using ice. Trawlers are equipped with plant to chill the sea water in vats.

Rennet
An extract from a calf's stomach. Can be produced from certain micro-organisms, eg *Mucor meihei* and *M. pusillus*.

Rennet contains mainly the *enzyme* chymosin (rennin) and some *pepsin*. It is responsible for coagulating milk during cheese-making, by breaking down the protective κ-casein (kappa) and allowing *calcium* and phosphate to gel with the *casein* (α_s).

Retinol
The name for vitamin A, or more specifically vitamin A_1 alcohol. *Carotenoids*, showing *pro-vitamin A* activity are converted to retinol in the body. An aldehyde version combines with a *protein* to produce visual purple needed for vision in poor light.

Deficiency causes night blindness, growth stunting and skin problems. Main sources are fish liver oils, liver, dairy products and margarine.

Retrogradation (of starch)
The reverse of gelatinization, in that water is expelled from the gel. The *starch* gradually undergoes a colloidal change and tends to contract with the loss of water. Staling of bread involves

retrogradation of starch and in this case can be reversed some-what by reheating the loaf.

Riboflavin
Vitamin B_2 which is combined with *proteins* to form part of the enzyme systems essential for oxidation of *carbohydrate* and the release of energy.

Deficiency improves oxidation carried out within the cell, whilst other symptoms include swollen and cracked lips and enlarged tongue. It is found in milk, eggs, liver and pulses. It is also a useful food colour giving attractive yellow shades (*E101*).

Ribonucleotides
Formed from the sugar ribose, phosphoric acid and a base such as guanine or inosine. They are *flavour enhancers* and operate in a similar manner to *MSG*, with which they can be blended.

They are found naturally in yeast, fish and meat. Commercially ribonucleotides are produced by micro-organisms.

Rickets
The disease caused by a deficiency of *vitamin D*. *Calcium* absorp-tion and bone formation is impaired leading to the bending of bones in the legs and other deformities. Vitamin D is found in fish liver oil and dairy products, and is sythesized in the skin as a result of sunlight.

Rigor mortis
The permanent contraction of animal muscle after death leading to rigidity in the carcass. The muscle *proteins*, actin and myosin, combine to produce actomyosin. The rigidity is lost after several hours due to the action of proteolytic *enzymes* breaking down the actomyosin.

Meat cooked whilst still in *rigor* is tougher and darker than that left until the rigidity has disappeared before cooking.

Ripening of cheese
Changes produced in 'green' cheese by *enzymes*, bacteria and moulds. Three general reactions are involved: fats are hydrolysed by proteolytic enzymes, and *amino acids* and *fatty acids* are released. These changes produce the flavour and texture of cheese; in well-matured cheese the reactions have been allowed to occur over a long period.

Ripening of fruit
Changes that occur after harvesting a mature fruit to produce a

product ripe for eating or processing. *Polysaccharides* such as *pectin* and *starch* are broken down to produce sugars, so a softer, sweeter fruit results. *Chlorophyll* is broken down to reveal other *pigments* such as red *carotenoids* or new ones are synthesized. The fruit becomes juicier and acidity is masked by the production of sugars.

Ripeness is a variable term depending on the end-use of the fruit. Fruit for jam-making is usually underripe, with a high *pectin* content. Fruit for juice-making is fully ripe with *polysaccharides* such as pectin, well hydrolyzed to sugars.

Roller drying see *Drying*

Roughage
A term used to mean dietary fibre, usually supplied as *cellulose*, *hemicellulose* and *pectin*. Basically roughage bulks up food and enables the intestines to function more easily.

Saccharin
An artificial sweetner, 550 times as sweet as *sucrose*. Chemically, *benzoic sulphimide*.

Salmonella
A genus of *gram negative* bacteria which are involved in many food poisoning outbreaks. They are found in many foods particularly poultry, meat and egg products. Typhoid, *S. typhi*, is a member of the genus.

Members of the genus are often given the name of the place where they were first discovered, eg *S. dublin*, *S. london*, *S. montevideo*. Some can survive refrigeration, but all are destroyed by adequate heating.

Screening
A method of size separation or sieving. Used for dry cleaning of foods and *sorting*.

'Sell-by' date
A requirement of the Food Labelling Regulations 1980. The last date of sale of the food, after which date it must be withdrawn from sale.

'Best before' date is the date up to which the food can normally be expected to retain its freshness.

Sequestrants
Substances which combine with metal ions. Examples include

citric acid and EDTA (ethylene diamine tetracetic acid). Traces of metals can initiate *oxidative rancidity* and so the use of a sequestrant will prevent this.

Single cell proteins
Proteins produced by bacteria or fungi growing on, or *fermenting*, suitable substrates such as waste cereals.

Size reduction
A requirement for some processes that food materials be made smaller. Crystalline substances are reduced by rollers or hammer mills. Fibrous materials need to be cut, diced or shredded.

Slow-set pectin see *Pectin*

Smoking
Meat products and fish may be smoked to aid preservation and give flavour. Smoke from burning hardwoods gives the best flavour. The preservative effect comes from *phenols, acids* and aldehydes in the smoke. Smoking also causes surface dehydration of the product. In 'hot smoking' the fish is cooked at the same time as being smoked.

Solvent extraction
Solvents, such as petroleum ether, are used to extract *oils* from vegetable sources, particularly seeds. The solvent is then distilled off to leave a crude oil which needs refining.

Sorting
Separation of food raw materials according to one characteristic such as size, weight or colour. Often confused with *grading* or quality separation.

Spray drying see *Drying*

Stabilizers
Substances, often complex *polysaccharides*, that have the ability to absorb considerable quantities of water. This property makes them good thickening agents, many being able to produce gels. Most can act as *emulsifiers* and prevent fat separation. Examples are *gums, cellulose* derivatives and gelatine.

Star marking (frozen products)
Introduced in 1964 for frozen products and freezers. One star means a temperature of $-6\,°C$ and gives storage of one week. Two stars indicate a temperature of $-12\,°C$ with up to 1 month of

storage. Three stars mean a temperature of $-18\,°C$ and up to three months of storage. Four stars on a freezer indicates an appliance capable of freezing food more rapidly, as well as storing frozen products.

Starch
A *polysaccharide* made from chains of α-glucose. Two forms exist: (a) *amylose* which is a straight chain in the form of a coil and (b) *amylopectin* which is a highly branched form.

Starch exists as granules in food products which are unique in appearance and size for each product.

- Phosphate cross-bonded starch
  starch treated with phosphoric acid to give the *amylose* fraction the appearance and properties of *amylopectin*. The result is a stable starch which does not retrograde (see *Retrogradation*) in canned or frozen products.

- Pre-gelatinized starches
  starch is mixed with water and heated to produce a gel which is then dried. The dried starch thickens instantly on the addition of cold water or milk. Used extensively in instant products, particularly desserts.

Starter cultures
Special cultures of bacteria incubated under ideal conditions to be added to foods to start *fermentation*, particularly in wines, bread, yoghurt and cheese.

Lactic acid bacteria, eg *Streptococcus* and *Lactobacillus* species are commonly used.

Sterilization
The achievement, usually by heating, of a complete absence of life (ie micro-organisms). In reality this is difficult to achieve as a few heat-resistant spores usually survive. The concept of *'commercial stability'* is acceptable to industry where some spores survive but do not usually germinate or grow.

Struvite
Small, sharp crystals found in canned fish and crustaceans. The crystals are ammonium magnesium phosphate. *Sequestrants* prevent their formation.

Sucrose
Common sugar, either from beet or cane. A *disaccharide* made from α-glucose and β-fructose. A non-*reducing sugar*.

Crude brown sugar is around 97 per cent sucrose, white sugar is almost pure sucrose at 99.9 per cent. Sucrose is usually given a sweetness value of 100 and all other sugars are compared with this.

Sulphur dioxide
A permitted *preservative* use in many products. Usually added as sulphite or metabisulphite. Also protects *vitamin C* but destroys *vitamin B$_1$*.
Used in fruit products, sausages, wines and campden tablets.

Surfactants
Basically *emulsifiers*. Lower the surface tension (usually of water) between two immiscible liquids in a product, thus aiding emulsification.

Syneresis
Loss of liquid from a gel on standing or as a result of damage.

Tainting
The transfer of odours from one food to another or from packaging materials and diesel fumes. The problem is difficult to trace, particularly in frozen foods during cold storage.

Tannins
Phenolic compounds of very complex structure. Responsible for the astringency of red wines, tea, coffee and apples. They also give body and fullness to the flavour of the product.

Tartar (cream of)
Potassium hydrogen tartrate. A weak acid used with sodium hydrogen carbonate (bicarbonate of soda) in traditional *baking powders*. Can be found as a precipitate in wine barrels after several months or years of storage.

Tartaric acid
An acid containing two acid groups (—COOH), occurring in some plants, particularly grapes. Used to increase acidity of jams, lemonade and desserts.

Tartrazine
A yellow to orange colour of complex structure, an azo dye. It has been implicated in allergic reactions and as a cause of hyperactivity in children. Has been withdrawn from most products, but is still permitted in many countries (*E102*).

197

Tempering
A process in chocolate manufacture aimed at achieving a proper consistency and hardness in the product. The cocoa butter has to be in a stable, crystal form. The process consists of melting the chocolate, heating it to 49 °C (120 °F), and cooling it with agitation down to 24 °C (84 °F).

Terpenoids
Isoprenoid derivatives made from two isoprene units ($C_5 H_8$). Found mainly in many naturally occurring *flavours*, particularly *essential oils*.

- monoterpenoids
  are based on two isoprene units ($C_{10} H_{16}$) and are alcohols, aldehydes and ketones derived from these (see *Monoterpenes*).
- diterpenoids
  are based on four isoprene units ($C_{20} H_{32}$) (see *Diterpenes*).

Tetrapyrrole derivatives
Composed of four pyrrole rings held together by methane bridges. The basic structure of *haemoglobin*, *myoglobin* and *chlorophyll*.

Textured vegetable protein (TVP)
Usually refers to soya *protein* which is produced from the soya bean and is defatted. The TVP is either extruded or spun into meat-like cubes. It must then be rehydrated before use. Nutritionally the product is good, but may be deficient in methionine.

Thiamine
Vitamin B_1 is needed as part of a coenzyme and deficiency results in the disease *beri-beri*. Symptoms include heart enlargement, cardiac failure, sensory and gastric disorders. The vitamin is found in yeast, eggs, pulses and meat.

Tocopherols
A group of eight compounds which show *vitamin E* activity. Found in many foods, particularly wheat germ and vegetable oils. Deficiency is unusual, leading to blindness and anaemia. Tocopherols are natural *antioxidants*.

Deficiency in rats causes sterility which probably has led to exaggerated claims for the use of vitamin E in humans.

Triglycerides
Made from *glycerol* and three *fatty acids*. Mixed triglycerides

contain three different fatty acids, simple triglycerides contain the same fatty acid. Mixed triglycerides are the basic constituents of natural fats.

Trimethylamine
The smell of bad fish. Produced from trimethylamine oxide by bacterial action in fish. Can be used as a measure of fish deterioration.

Trimethylxanthine see *Caffeine*

Ultra-high temperature UHT
High temperature, short time *sterilization* of milk usually carried out in plate heat exchangers. UHT milk is usually called 'Long-Life' milk.

Uperisation
A *sterilization* method employing direct injection of steam under pressure. Extensively used in Europe for producing *UHT* milk.

Uronic acids
Derivatives of sugars, with a carboxyl group (—COOH) at carbon atom 6 position. Examples, *galacturonic acid* (from *galactose*) which is a component of *pectin*.

Vitamins
Organic substances required in small quantities but cannot by synthesized by the body. Vitamins absent from, or low in the diet may result in deficiency disease which may be fatal if not remedied.

Vitamin A   see *Retinol*
Vitamin B_1 see *Thiamine*
Vitamin B_2 see *Riboflavin*
Vitamin B_6 see *Pyridoxine*
Vitamin B_{12} see *Cyanocobalanin*
Vitamin C   see *Ascorbic acid*
Vitamin D   see *Calciferols*
Vitamin E   see *Tocopherols*
Vitamin K   see *Naphthoquinones*

Votator
A double tubular heat-exchanger consisting of one tube within another. The centre tube is usually chilled by a circulatory chilled brine. Used mainly in margarine manufacture to ensure emulsification and texture development.

Water activity (a$_w$)

The amount of water in a food can be described in terms of water activity (a$_w$). Levels of moisture are compared with pure water which has an a$_w$ of 1.0. Preservation methods such as dehydration, concentration, the addition of salt and sugar, rely on lowering the water activity.

At a$_w$ levels of 0.9 and above most bacteria grow readily. At levels of 0.85–0.9 yeasts tend to dominate. Moulds grow in the range 0.80–0.85. Halophilic bacteria (like salt) grow down at 0.75 and osmophilic yeasts (like sugar) at 0.6.

Waxes

Complex fatty materials, usually containing alcohols with long chains of carbon atoms. Found on the surface of some fruits, eg apples.

Wheat proteins

Known collectively as *gluten*. A blend of two main protein types gliadins (40–50 per cent), glutenins (40–50 per cent) with small amounts of albumin, globulin and proteose. Responsible for the elasticity and extensibility of dough during bread-making.

Whey proteins

Proteins from milk after the removal of *casein*. Mainly comprise of α-lactalbumin and β-lactoglobulin. Can be concentrated or dried and used in foods for their functional properties such as: foaming ability, emulsification, and texture modification.

Xanthophylls

Carotenoid pigments that are yellow. They must contain an hydroxyl group (—OH). For example, cryptoxanthin, the chief *pigment* of maize, paprika and mandarin orange.

Xylose

A pentose sugar (5 carbon atoms) found in plants, a main constituent of hemicelluloses.

Yeasts

Fungi which are usually involved in fermentation and spoilage of sweetened or salted products.

Saccharomyces cerevisiae – used for bread and beer making.
S. ellipsoideus – used in wine making.

Yeasts are useful sources of *protein*, and B *vitamins*. Yeast extracts and hydrolysates are used as savoury flavours of soups and meat products.

Yolk index (of eggs)
An index of freshness for eggs. It is the ratio of the height of yolk to the diameter. The index decreases as the egg ages and deteriorates.

Zwitterion
An *amino acid* at the iso-electric point, the pH at which the net charge on the amino acid is zero. Usually one positive charge on the amino group is balanced by one negative charge on the carboxyl group. The solubility, viscosity and electrical conductivity of the amino acid are at a minimum in this form.

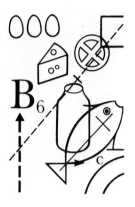

Appendix

This Appendix comprises a complete list of given permitted additives, their uses and E numbers. It is based on a list to be found in the Ministry of Agriculture, Fisheries and Foods' booklet *Food Additives – the Balanced Approach* (HMSO, London, © Crown Copyright 1987).

Note: Not all permitted additives are in use

Complete list of permitted additives

Colours

Number	Name	Products in which found
E100	curcumin	flour confectionery, margarine, rice
E101	riboflavin	sauces, processed cheese
101(a)	riboflavin-5'-phosphate	preserves
E102	tartrazine	soft drinks (formerly in many yellow/orange products)
E104	quinoline yellow	Scotch eggs
107	yellow 2G	
E110	sunset yellow FCF	biscuits, soups, Swiss rolls
E120	chochineal	alcoholic drinks
E122	carmoisine	jams, preserves, soups, sweets
E123	amaranth	cake mixes, soups
E124	ponceau 4R	dessert mixes, soups
E127	erythrosine	glacé cherries, trifle mix
128	red 2G	sausages
E131	patent blue V	Scotch eggs
E132	indigo carmine	biscuits, sweets
133	brilliant blue FCF	canned vegetables
E140	chlorophyll	soap, oils
E141	copper complexes of chlorophyll and chlorophyllins	green vegetables in liquid
E142	green S	pastilles, tinned peas
E150	caramel	beer, soft drinks, sauces, gravy browning
E151	black PN	sauces
E153	carbon black (vegetable carbon)	liquorice, preserves
154	brown FK	kippers (will be deleted when suitable replacement available)
155	brown HT (chocolate brown HT)	chocolate cake
160(a)	alpha-carotene; beta-carotene; gamma-carotene	margarine, soft drinks
E160(b)	annatto; bixin; norbixin	crisps, butter, coleslaw
E160(c)	capsanthin; capsorubin	cheese slices
E160(d)	lycopene	
E160(e)	beta-apo-8'-carotenal	
E160(f)	ethyl ester of beta-apo-8'-carotenoic acid	
E161(a)	flavoxanthin	
E161(b)	lutein	
E161(c)	cryptoxanthin	
E161(d)	rubixanthin	
E161(e)	violaxanthin	
E161(f)	rhodoxanthin	
E161(g)	canthaxanthin	biscuits (likely to be deleted)
E162	beetroot red (betanin)	ice-cream, liquorice
E163	anthocyanins	yoghurts
E171	titanium dioxide	sweets, horseradish sauce
E172	iron oxides; iron hydroxides	dessert mixes, cake mixes
E173	aluminium	
E174	silver	sugar, confectionery, cake decorations
E175	gold	

Colours (cont)

Number	Name	Products in which found
E180	pigment rubine (lithol rubine BK)	rind of cheese
	methyl violet	surface marking of citrus fruit
	paprika	canned vegetables
	saffron; crocin	
	sandalwood; santolin	
	turmeric	soups

Preservatives

Number	Name	Products in which found
E200	sorbic acid	soft drinks, fruit yoghurts, processed cheese slices, sweets
E201	sodium sorbate	
E202	potassium sorbate	pizza, cheese spread, cakes
E203	calcium sorbate	
E210	benzoic acid	
E211	sodium benzoate	
E212	potassium benzoate	
E213	calcium benzoate	
E214	ethyl 4-hydroxybenzoate (ethyl para-hydroxybenzoate)	
E215	ethyl 4-hydroxybenzoate, sodium salt (sodium ethyl para-hydroxybenzoate)	pickles, cheesecake mix, sauces, flavourings, beer, jam, salad cream, soft drinks, fruit pulp, fruit-based pie fillings, marinated herring and mackerel, desserts
E216	propyl 4-hydroxybenzoate (propyl para-hydroxybenzoate)	
E217	propyl 4-hydroxybenzoate, sodium salt (sodium propyl para-hydroxybenzoate)	
E218	methyl 4-hydroxybenzoate (methyl para-hydroxybenzoate)	
E219	methyl 4-hydroxybenzoate, sodium salt (sodium methyl para-hydroxybenzoate)	
E220	sulphur dioxide	
E221	sodium sulphite	
E222	sodium hydrogen sulphite (sodium bisulphite)	dried fruit, dehydrated vegetables, fruit juices, fruit pulp, fruit syrups, drinks, wine, egg yolk, pickles
E223	sodium metabisulphite	
E224	postassium metabisulphite	
E226	calcium sulphite	
E227	calcium hydrogen sulphite (calcium bisulphite)	
E230	biphenyl (diphenyl)	
E231	2-hydroxybiophenyl (orthophenylphenate)	
E232	sodium biphenyl-2-yl oxide (sodium orthophenylphenate)	fruit skins, surface of citrus fruits, surface treatment of bananas
E233	2-(thiazol-4-yl) benzimidazole (thiabendazole)	
234	nisin	cheese, clotted cream, canned foods
E239	hexamine (hexamethylenetetramine)	marinated herring and mackerel
E249	potassium nitrite	
E250	sodium nitrite	cooked meats, bacon, ham, cured meats, corned beef,
E251	sodium nitrate	tongue, some cheeses
E252	potassium nitrate	
E280	propionic acid	
E281	sodium propionate	bread, flour confectionery, dairy products, pizza,
E282	calcium propionate	Christmas pudding
E283	potassium propionate	

Antioxidants

Number	Name	Products in which found
E300	L-ascorbic acid	
E301	sodium L-ascorbate	
E302	calcium L-ascorbate	fruit drinks, also used to improve flour and bread dough,
E304	6-0-palmitoyl-L-ascorbic acid (ascorbyl palmitate)	dried potato, sausages, meat loaf, Scotch eggs, steak cubes
E306	extracts of natural origin rich in tocopherols	
E307	synthetic alpha-tocopherol	vegetable oils, cereal-based baby foods
E308	synthetic gamma-tocopherol	
E309	synthetic delta-tocopherol	
E310	propyl gallate	
E311	octyl gallate	vegetable oils, chewing gum, margarine
E312	dodecyl gallate	
E320	butylated hydroxyanisole (BHA)	beef stock cubes, cheese spread, biscuits, convenience foods
E321	butylated hydroxytoluene (BHT)	chewing gum
E322	lecithins	low fat spreads, in chocolate as an emulsifer, confectionery
	diphenylamine ethoxyquin	on apples and pears used to prevent 'scald' (discolouration)

Emulsifiers and Stabilizers

Number	Name	Products in which found
E400	alginic acid	
E401	sodium alginate	
E402	potassium alginate	ice-cream, soft and processed cheese, cake mixes, salad
E403	ammonium alginate	dressings, cottage cheese, synthetic cream
E404	calcium alginate	
E405	propane-1, 2-diol alginate (propylene glycol alginate)	
E406	agar	ice-cream, frozen trifle
E407	carrageenan	quick setting jelly mixes, milk shakes
E410	locust bean gum (carob gum)	salad cream
E412	guar gum	packet soups, meringue mixes, sauces
E413	tragacanth	salad dressings, processed cheese, cream cheese
E414	gum arabic (acacia)	confectionery
E415	xanthan gum	sweet pickle, coleslaw, horseradish cream
416	karaya gum	soft cheese, brown sauce
430	polyoxyethylene (8) stearate	
431	polyoxyethylene (40) stearate	
432	polyoxyethylene (20) sorbitan monolaurate (Polysorbate 20)	
433	polyoxyethylene (20) sorbitan mono-oleate (Polysorbate 80)	
434	polyoxyethylene (20) sorbitan monopalmitate (Polysorbate 40)	bakery products, confectionery creams, cakes
435	polyoxyethylene (20) sorbitan monostearate (Polysorbate 60)	
436	polyoxyethylene (20) sorbitan tristearate (Polysorbate 65)	
E440(a)	pectin	
E440(b)	amidated pectin	jams, preserves, desserts
	pectin extract	
442	ammonium phosphatides	cocoa and chocolate products

Emulsifiers and Stabilizers (cont)

Number	Name	Products in which found
E460	microcrystalline cellulose; alpha-cellulose (powdered cellulose)	high-fibre bread, grated cheese, low fat spreads, edible ices, gâteaux
E461	methylcellulose	
E463	hydroxypropylcellulose	
E464	hydroxypropylmethylcellulose	
E465	ethylmethylcellulose	
E466	carboxymethylcellulose, sodium salt (CMC)	jelly, pie filling
E470	sodium, potassium and calcium salts of fatty acids	cake mixes
E471	mono- and di-glycerides of fatty acids	
E472(a)	acetic acid esters of mono- and di-glycerides of fatty acids	
E472(b)	lactic acid esters of mono- and di-glycerides of fatty acids	frozen desserts, dessert toppings, mousse mixes, continental sausages, bread, frozen pizza
E472(c)	citric acid esters of mono- and di-glycerides of fatty acids	
E472(e)	mono- and diacetyltartaric acid esters of mono- and di-glycerides of fatty acids	
E473	sucrose esters of fatty acids	coatings for fruit
E474	sucroglycerides	edible ices
E475	polyglycerol esters of fatty acids	cakes and gâteaux
476	polyglycerol esters of polycondensed fatty acids of castor oil (polyglycerol polyricinoleate)	chocolate-flavour coatings for cakes
E477	propane-1, 2-diol esters of fatty acids	instant desserts
478	lactylated fatty acid esters of glycerol and propane-1, 2-diol	
E481	sodium stearoyl-2-lactylate	bread, cakes, biscuits, gravy granules
E482	calcium stearoyl-2-lactylate	
E483	stearyl tartrate	
491	sorbitan monostearate	
492	sorbitan tristearate	cake mixes
493	sorbitan monolaurate	
494	sorbitan mono-oleate	
495	sorbitan monopalmitate	
	dioctyl sodium sulphosuccinate	used in sugar refining to help crystallization
	extract of quillaia	used in soft drinks to promote foam
	oxidatively polymerised soya bean oil	emulsions used to grease bakery tins
	polyglycerol esters of dimerised fatty acids of soya bean oil	

Sweeteners

Number	Name	Products in which found
	acesulfame potassium	canned foods, soft drinks, table-top sweeteners
	aspartame	soft drinks, yoghurts, dessert and drink mixes, sweetening tablets
	hydrogenated glucose syrup	
	isomalt	
E421	mannitol	sugar-free confectionery
	saccharin	
	sodium saccharin	soft drinks, cider, sweetening tablets
	calcium saccharin	
E420	sorbitol; sorbitol syrup	sugar-free confectionery, jams for diabetics
	thaumatin	table-top sweeteners, yoghurt
	xylitol	sugar-free chewing gum

Others

Number	Name	Products	Use
E170	calcium carbonate		base, firming or release agent
E260	acetic acid		
E261	potassium acetate		
E262	sodium hydrogen diacetate	pickles, crisps, salad cream, bread	acid/acidity regulators
262	sodium acetate		
E263	calcium acetate	quick set jelly mix	firming agent
E270	lactic acid	salad dressing, margarine	acid, antifungal action
E290	carbon dioxide	fizzy drinks	carbonating agent, packaging gas
296	DL-malic acid; L-malic acid	low calorie squash, soup	acid
297	fumaric acid	soft drinks, sweets, biscuits, dessert mixes, pie fillings	acid
E325	sodium lactate	jams, preserves, sweets, flour confectionery	buffer/humectant
E326	potassium lactate	jams, preserves, jellies,	buffer/humectant
E327	calcium lactate	canned fruit, fruit pie filling	
E330	citric acid	many products	acid
E331	sodium dihydrogen citrate (monosodium citrate); disodium citrate; trisodium citrate	sweets, gâteaux mixes, soft drinks, jams, preserves, sweets, processed cheese, canned fruit, dessert mixes	acid/flavour buffer, sequestrants, calcium salts are firming agents
E332	potassium dihydrogen citrate (monopotassium citrate); tripotassium citrate		
E333	monocalcium citrate; dicalcium citrate; tricalcium citrate		
E334	L-(+)-tartaric acid	confectionery, drinks, preserves, meringue pie mix, soft drinks, biscuit creams and fillings, dessert mixes, processed cheese	acid/flavouring, buffer, emulsifying salts, sequestrants
E335	monosodium L-(+)-tartrate; disodium L-(+)-tartrate		
E336	monopotassium L-(+)-tartrate (cream of tartare); dipotassium L-(+)-tartrate		
E337	potassium sodium L-(+)-tartrate		
E338	orthophosphoric acid (phosphoric acid)	soft drinks, cocoa, dessert mixes, non-dairy creamers, processed cheese	acid/flavouring, sequestrants, emulsifying agents, buffers
E339	sodium dihydrogen orthophosphate; disodium hydrogen orthophosphate; trisodium orthophosphate		
E340	potassium dihydrogen orthosphosphate; dipotassium hydrogen orthophosphate; tripotassium orthophosphate		
E341	calcium tetrahydrogen diorthophosphate; calcium hydrogen orthophosphate; tricalcium diorthophosphate		
350	sodium malate, sodium hydrogen malate	jams, sweets, cakes, biscuits, processed fruit and vegetables	buffers, humectants, calcium salts are firming agents in canned fruit and vegetables
351	potassium malate		
352	calcium malate, calcium hydrogen malate		
353	metatartaric acid	wine	sequestrant
355	adipic acid	sweets, synthetic cream desserts	buffer/flavouring
363	succinic acid	dry food and beverage mixes	buffer/flavouring
370	1,4-heptonolactone	dried soups, instant desserts	acid, sequestrant
375	nicotinic acid	bread, flour, breakfast cereals	colour stabilizer, nutrient

Others (cont)

Number	Name	Products	Use
380	triammonium citrate	processed cheese	buffer, emulsifying salt
381	ammonium ferric citrate	bread	iron supplement
385	calcium disordium ethylenediamine-NNN′N′-tetra-acetate (calcium disodium EDTA)	canned shellfish	sequestrant
E422	glycerol	cake icing, confectionery	humectant, solvent
E450(a)	disodium dihydrogen diphosphate, trisodium diphosphate, tetrasodium diphosphate, tetrapotassium diphosphate	whipping cream, meat products, bread, processed cheese, canned vegetables	buffers, sequestrants, emulsifying salts, stabilisers, raising agents
E450(b)	pentasodium triphosphate, pentapotassium triphosphate		
E450(c)	sodium polyphosphates, potassium polyphosphates		
500	sodium carbonate, sodium hydrogen carbonate (bicarbonate of soda), sodium sesquicarbonate	jams, jellies, self-raising flour, wine, cocoa, biscuits, icing sugar	bases, aerating agents, anti-caking agents
501	potassium carbonate, potassium hydrogen carbonate		
503	ammonium carbonate; ammonium hydrogen carbonate		
504	magnesium carbonate		
507	hydrochloric acid	tomato juice	processing aid
508	potassium chloride	table salt replacement	gelling agent, salt substitute
509	calcium chloride	canned fruit and vegetables	firming agent
510	ammonium chloride	bread	yeast food
513	sulphuric acid		
514	sodium sulphate	colours	diluent
515	potassium sulphate	salt	salt substitute
516	calcium sulphate	bread	firming agent, yeast food
518	magnesium sulphate	bread	firming agent
524	sodium hydroxide	cocoa, jams, sweets	base
525	potassium hydroxide	sweets	base
526	calcium hydroxide	sweets	firming agent, neutralizing agent
527	ammonium hydroxide	cocoa, food colouring	diluent and solvent for food colours, base
528	magnesium hydroxide	sweets	base
529	calcium oxide	sweets	base
530	magnesium oxide	cocoa products	anti-caking agent
535	sodium ferrocyanide	salt, wine	anti-caking agent
536	potassium ferrocyanide		
540	dicalcium diphosphate	cheese	buffer, neutralizing agent
541	sodium aluminium phosphate	cake mixes, self-raising flour, biscuits	acid, raising agent
542	edible bone phosphate		anti-caking agent
544	calcium polyphosphates	processed cheese	emulsifying salt
545	ammonium polyphosphates	frozen chicken	emulsifier, texturizer
551	silicon dioxide (silica)	skimmed milk powder, sweeteners	anti-caking agent
552	calcium silicate	icing sugar, sweets	anti-caking agent, release agent
553(a)	magnesium silicate synthetic; magnesium trisilicate	sugar confectionery	anti-caking agent
553(b)	talc	tabletted confectionery	release agent
554	aluminium sodium silicate	packet noodles	anti-caking agent

Others (cont)

Number	Name	Products	Use
556	aluminium calcium silicate		anti-caking agent
558	bentonite		anti-caking agent
559	kaolin		anti-caking agent
570	stearic acid		anti-caking agent
572	magnesium stearate	confectionery	emulsifier, release agent
575	D-glucono-1,5-lactone (glucono delta-lactone)	cake mixes, continental sausages	acid, sequestrant
576	sodium gluconate		
577	potassium gluconate	dietary supplement, jams, dessert mixes	sequestrants, buffer, firming agent
578	calcium gluconate		
620	L-glutamic acid		
621	sodium hydrogen L-glutamate (monosodium glutamate; MSG)		
622	potassium hydrogen L-glutamate (monopotassium glutamate)		
623	calcium dihydrogen di-L-glutamate (calcium glutamate)	savoury foods and snacks, soups, sauces, meat products	flavour enhancers
627	guanosine 5'-disodium phosphate (sodium guanylate)		
631	inosine 5'-disodium phosphate (sodium inosinate)		
635	sodium 5'-ribonucleotide		
636	maltol	cakes and biscuits	flavourings and flavour enhancers
637	ethyl maltol		
900	dimethylpolysiloxane		anti-foaming agent
901	beeswax	sugar, chocolate confectionery	glazing agent
903	carnauba wax	sugar, chocolate confectionery	glazing agent
904	shellac	apples	waxing agent
905	mineral hydrocarbons	dried fruit	glazing/coating agent
907	refined microcrystalline wax	chewing gum	release agent
920	L-cysteine hydrochloride		
924	potassium bromate		flour treatment agents to improve texture
925	chlorine	bread, cake, biscuit dough	
926	chlorine dioxide		
927	azodicarbonamide		
	aluminium potassium sulphate	chocolate-coated cherries	firming agent
	2-aminoethanol	caustic lye used to peel vegetables	base
	ammonium dihydrogen orthophosphate, diammonium hydrogen othophosphate	yeast food	buffer
	ammonium sulphate	yeast food	
	benzoyl peroxide	flour	bleaching agent
	butyl stearate		release agent
	calcium heptonate	prepared fruit and vegetables	firming agent, sequestrant
	calcium phytate	wine	sequestrant
	dichlorodifluoromethane	frozen food	propellant and liquid freezant
	diethyl ether		solvent
	disodium dihydrogen ethylenediamine-NNN'N'-tetra-acetate (disodium dihydrogen EDTA)	brandy	sequestrant

211

Others (cont)

Number	Name	Products	Use
	ethanol (ethyl alcohol)		
	ethyl acetate		
	glycerol mono-acetate (monoacetin)	food colours, flavourings	solvent, dilutent
	glycerol di-acetate (diacetin)		
	glycerol tri-acetate (triacetin)		
	glycine		sequestrant, buffer, nutrient
	hydrogen		packaging gas
	nitrogen		packaging gas
	nitrous oxide	whipped cream	propellant in aerosol packs
	octadecylammonium acetate	yeast foods in bread	anti-caking agent
	oxygen		packaging gas
	oxystearin	salad cream	sequestrant, fat crystallization inhibitor
	polydextrose	reduced and low calorie foods	bulking agent
	propan-1,2-diol (propylene glycol)	colourings and flavourings	solvent
	propan-2-ol (isopropyl alcohol)		
	sodium helptonate	edible oils	sequestrant
	spermaceti		release agent
	sperm oil		release agent

Index

Please also refer to the Glossary (pages 159–201)